ELEMENTARY
SYSTEMS
MATHEMATICS

McGRAW-HILL
BOOK COMPANY
New York
St. Louis
San Francisco
Auckland
Düsseldorf
Johannesburg
Kuala Lumpur
London
Mexico
Montreal
New Delhi
Panama
Paris
São Paulo
Singapore
Sydney
Tokyo
Toronto

ROBERT E. MACHOL

Professor of Systems
Graduate School of Management
Northwestern University

Elementary Systems Mathematics

LINEAR PROGRAMMING FOR BUSINESS AND THE SOCIAL SCIENCES

This book was set in Times Roman by Textbook Services, Inc.
The editors were B. J. Clark and James W. Bradley;
the cover was designed by Chris Kristiansen;
the production supervisor was Leroy A. Young.
The drawings were done by B. Handelman Associates, Inc.
Kingsport Press, Inc., was printer and binder.

Library of Congress Cataloging in Publication Data

Machol, Robert Engel, date
Elementary systems mathematics.

Includes index.
1. Linear programming. I. Title.
T57.74.M3 519.7'2 75-34014
ISBN 0-07-039373-7

ELEMENTARY
SYSTEMS
MATHEMATICS

LINEAR PROGRAMMING FOR BUSINESS
AND THE SOCIAL SCIENCES

1 2 3 4 5 6 7 8 9 0 KPKP 7 8 3 2 1 0 9 8 7 6

To FLORENCE, MARGOT, and KEN

CONTENTS

TO THE STUDENT

This is an intensely mathematical book, and you, the reader, are not a mathematician, but don't panic. It is assumed that many of you have never had a mathematics course beyond that taught in the usual junior year of high school; but in that case, you should have understood that material reasonably well. Others of you will have reviewed algebra in college. But it is not assumed that you have had fancy things like trigonometry, advanced algebra, or calculus. Of course if you have taken such courses, that is all to the good. If you've learned to program a computer, that too will help—not because any knowledge of computers is assumed, but because the kind of thinking required is similar to what is needed in this book. For the same reason, after you finish this book, you will find it fairly easy to learn to write computer programs.

If you do have minimum preparation, you'll have to work rather hard, especially in Chapter 1. Nonetheless, this book has been mastered by many students whose mathematical ability and training are no better than yours. Furthermore, we assure you that you do not have to learn any new mathematics, only new methods of applying the mathematics you already know.

As a matter of fact, you have probably been misled as to what mathematics is all about. It is not about arithmetic and algebra and esoteric formulas, although these are all often useful in mathematics. What mathematics is really about is relationships of any kind when these can be expressed quantitatively. For example, in a typical linear programming situation, we will consider the decision problem faced by a manufacturer who has to decide which of 37 different products he should produce, and how much of each, when he is subjected to many constraints involving marketing, availability of resources and production factilities, and so forth. We will find in many such cases that there exists a unique solution which is better than all others and that we can determine this solution numerically. But, perhaps more important, from our understanding of the mathematical properties which such solutions must have, we will be able to gain enormous insights. For example, we will know even before we solve the problem that some of the products should not be made and that some of the resources will not be fully utilized, and we will be able to deduce all sorts of peripheral information, such as the value of adding additional production facilities.

All this comes later. For the moment we wish to advise you how to use this book. Since it is a mathematics text, it cannot be read like a novel (unless you are unusually bright and well prepared). It should preferably be read at a desk, with scratch paper and pencils in front of you so that you may check each statement as it is made in the text. You may prefer to read the entire chapter once over lightly and then go through it carefully, checking as you go. The best strategy is to assume always that the book is wrong until you have done the calculation yourself and proved it right. If you do think you have found an error, we would be pleased to hear from you. Incidentally, we hope you will pardon the use of the "editorial we"; it seemed easier than writing "I," and sometimes it includes the reader as well as the author.

A man named J. D. Williams once wrote* that "it is customary in scientific writing to state the important points just as often as the unimportant ones; exactly once; and to assume that this vital information is graven in the minds of all readers. This is one of the wonders of science, for it is economical of space in the journals and it encourages researchers to write up their findings. It also ensures a certain permanence to the work, for it will be years before it is completely understood." This description may apply to most mathematics books, but it does not apply to this one. We have stated the important points several times. We hope our frequent use of "remember that" or "you will recall" does not bore you.

Because this is a mathematical book, everything is numbered in a highly organized fashion. We start with a Chapter 0, which is almost beneath your notice, but you should go through it to make sure you know everything in it. In each chapter the sections are numbered consecutively; for example, Chapter 1 has Sections 1.1, 1.2, and so forth. Some of the sections have subsections, which are numbered with an additional digit; for example under Section 1.3 we

*From "The Compleat Strategyst" by J. D. Williams. Copyright 1966. Used with permission of McGraw-Hill Book Company.

have subsections 1.3.1 and 1.3.2. The figures in each chapter are also numbered consecutively; in Chapter 1 we have Figures 1-1, 1-2, and so forth. We will frequently wish to refer to an equation or set of equations; we don't bother numbering an equation if we don't wish to refer to it again, but when we do they are numbered consecutively. For example, in Chapter 1 we have equations (1-1), (1-2), and so forth—note the parentheses! Finally, at the end of Chapter 1 you will find Problems 1-1, 1-2, and so forth. Sometimes we will refer to a "problem (4-6)"; the parentheses tell you that this is not a problem at the end of Chapter 4 but a set of equations in the middle of Chapter 4 which represents a particular problem. Thus it will help if you understand these conventions. To make up for all the notational devices you must learn, we have eliminated all abbreviations; we even resisted the temptation to use LP for linear programming.

There are no footnotes in this book (except as required by copyright laws), since we think that you would find them distracting. However, occasionally we could not resist introducing some sidelight or advanced concept that was not essential. These are printed in smaller type to warn you that they should be given the once-over-lightly treatment; that is, read them, but don't worry about them. Other than the small-type sections, if there is something you do not understand, go back over it until you do understand it, or get your instructor (or a fellow student) to explain it to you.

The problems will constitute the most important part of your learning, and you will probably find, when you try the problems, that you will have to go back and read the text once again. Even if you are a genius, you will not understand the later chapters unless you have mastered the earlier ones, and you cannot master them without the painful process of working problems. After many problems you will see, "This problem is continued in Problem...," and you should save these because you will need them when you work the subsequent problem. What's happening is that as you learn more about linear programming, you will be able to get deeper insights into what first seemed an elementary problem, and therefore to learn more about the solution.

Above all, do not be rebuffed by the symbolism. The mathematical notation which is used throughout this book has been developed over many generations and has proved itself to be not only concise and unambiguous but a real aid to understanding. Look for example, at the problem (2-1), presented in the six lines of equations on page 54. This must appear forbidding to you and absolutely incomprehensible. By the time you get to it, it will appear merely difficult. By the time you finish Chapter 2, it will appear easy. By the time you finish the book, it will appear absolutely trivial.

Have faith. And good luck!

ROBERT E. MACHOL

TO THE INSTRUCTOR

This book is intended for that vast group of undergraduates majoring in humanities or social sciences, or for graduate students in such fields as management, who wish to take one or two math courses by way of "distribution." Formerly such students were required to study calculus; more recently an option has been "finite mathematics," a potpourri of mathematical techniques not involving the infinitesimal calculus. Such courses provide only techniques which the student will not use again and will soon forget. What the student needs, and what this text provides, is a feeling for what mathematics is all about and an understanding of why it is such a useful and powerful way of looking at the world, in the context of a set of realistic and practical problems which he or she can learn to solve.

In the 20 years since the publication of the first book on linear programming, about 100 texts have appeared on this subject, but none is comparable to this work, where teaching linear programming itself is a major, but nonetheless secondary, objective. Some of the introductory texts, such as those by Hadley and Dantzig, are excellent but are hopelessly difficult for many students. At the other extreme are books which attempt to "teach all the necessary mathematics," but which leave the student discouraged and confused with their inverses, adjoints, and cofactors before he or she ever gets to linear programming.

No mathematical training beyond junior-level high school algebra is required, although the student is expected to be reasonably proficient with that material. While there is an introductory review of the elements of arithmetic and algebra, students who have never really learned high school algebra may need a quarter or semester of such material before starting this book. Of course if they have studied more advanced mathematics (trigonometry, calculus, advanced algebra) or have had a course in computers, so much the better. They probably do not intend ever to take another math course and will breathe a sigh of relief when this is over. But by then they will have learned to formulate and solve linear programming problems, as well as something about mathematics.

The instructor will share with the author impatience at doing certain things periphrastically when they could be done simply and straightforwardly by a matrix manipulation, but a delicate balance has been sought here. For example, "main diagonal" has obvious advantages over "northwest-to-southeast diagonal," but it is an extra word for the students to learn; and since we use it only once or twice, it seemed best not to burden them. On the other hand, to call a permutation matrix an identity matrix when it is not, as is done in many texts, is simply wrong, and I have tried to avoid such errors.

We use the notation of matrices, but none of the theory of linear vector spaces. In my experience, you can teach such students the algorithm for inverting a matrix or for computing a determinant, but you cannot really teach them what an inverse or a determinant is all about in the time available. Of course, at each simplex iteration the students are really computing the inverse of the basis matrix, but it seems best not to tell them (until near the end of the book), in accordance with optimizing the utilization of their intellectual resources toward the objective of comprehension of the methods and capabilities of mathematics. Not mathematical technique—that is not the issue. What distinguishes the mathematically mature student is not his or her ability to differentiate a polynomial or invert a matrix; it is the ability to define appropriate variables, write equations and inequalities that describe a situation, and read with understanding a text which uses symbolic descriptions. After all, the great bulk of these students will never actually use the simplex algorithm. We hope that they will gain some insights into model building, optimization, quantitative description, and the systems viewpoint. And we also hope they will learn that, by studying the properties which a solution must have, and the underlying theory (in this case duality), they can devise methods for determining the solution and also gain great insights into things like sensitivity. Such understanding is a significant aspect of mathematical maturity.

The book differs from other books in many other ways. The theory of duality permeates the entire book and is introduced before the student sees the simplex algorithm. Consequently, it is natural to introduce the dual simplex method early as a method of solving minimization problems and to delay the introduction of artificial variables (which students usually find difficult). The organization of the book is also different. It starts with the assignment problem, which is the simplest of all linear programming problems. Hence, the student is able to solve real optimization problems before becoming involved with the arithmetic and algebraic complexities of simplex. It then goes on to the transportation problem, which is more complicated, and rises finally to the general linear programming problem, at which point simplex is introduced. Sensitivity analysis is first introduced in the transportaton problem, where it is much easier to comprehend. Degeneracy is introduced early, as something fundamental rather than as a special trick: Degeneracy and alternate-optimal solutions are dual to one another, and the reason we use a dual algorithm on the assignment problem is because alternate optima are easier to deal with than degeneracies. This and other properties of the assignment and transportation problems and their algorithms are investigated in considerable depth after linear programming and the theory of duality have been developed so that the student finally obtains a considerable understanding of these two problems, which are by no means as uncomplicated as the usual presentations imply. This kind of sophistication—a deep understanding of at least one small area—is at the heart of scholarship.

The problems at the end of each chapter are, of course, of the utmost importance. I have personally written and worked every problem, adjusting the numbers to emphasize the desired points without unnecessarily messy arithmetic. The problems are mostly very different from one another (as distinguished from, say, the same old product-mix problem verbally disguised in different contexts or with different numbers). They have been arranged in several sections to make it easier for you to choose assignments. The first section contains the essential problems, which constitute the usual assignment for the chapter. The second section contains supplementary problems that may be used for drill if you feel that your students need them. "Extra drill problems" are also supplied, primarily for self-assignment by students who want more practice in mechanics. The last section contains more advanced problems which include optional extra concepts. You may wish to supplement the regular problems with drill problems for the weaker students and advanced problems for the stronger. For some classes I have promised (as an incentive) that the final exam will include one unassigned problem from the book. A special section of formulation problems is included in Chapter 4. They are particularly recommended, because formulation is most important and, for many readers, most difficult.

The instructor's manual is an integral part of this book and includes bits of instruction in connection with the answers to problems. With some groups of students I have found success in making the answer book available to them, ensuring, of course, that they have tried the problem themselves before they consult it and that they do not copy it. Alternatively, this manual contains hints which the instructor may use in going over the problem in the next day's lecture.

For students with minimal preparation, Chapter 1 is troublesome, and they need some help from the instructor to be comfortable with the c_j's and x_{ij}'s and Σ's, to understand algorithms, and, most important, to know what a variable is. The usual one-quarter or one-semester course would cover at least the Introduction through Chapter 6. It is intended that the material through Chapter 6 be read in the order presented and without omissions. The sections of Chapter 7 may be read in any order, but perhaps the order given is best, and one would hope that the students could get at least through Section 7.2 (I have covered this much in a 10-week quarter). I also hope that they would read the conclusion (Chapter 8) even if they did not finish Chapter 7.

No bibliography is given. The usual references to the original work of Dantzig, Kuhn, Koopmans, and others, are fatuous for the readers of this book. The older literature is excellently documented in Dantzig's "Linear Programming and Extensions" (1964); the more recent literature can be traced from articles in such journals as *Operations Research, Mathematical Programming,* and *Mathematics of Operations Research.*

Professors Frederic Murphy and David Peterson were kind enough to help with an early version of this book, and Professor Paul Gray read the final manuscript very carefully. I taught from a typescript version three times over a period of two years, and feedback from my students at the Graduate School of Management of Northwestern was most valuable. I particularly wish to thank Ms. Theresa Bonk for TLC exercised in typing the manuscript many times.

ROBERT E. MACHOL

ELEMENTARY
SYSTEMS
MATHEMATICS

0

REVIEW OF ELEMENTARY MATHEMATICS

The subject matter of this book is the application of quantitative methods (that is, mathematics) to decision making. No advanced mathemetics is required, and the student who has mastered the mathematics curriculum through the material normally taught in the junior year of high school will be able to follow the entire book. However, you, the reader, may have been exposed to this material several years ago, and so it is the purpose of this chapter to review the mathematical prerequisites. It is essential that the material in this chapter and the next be thoroughly understood, not merely on the level of "I understand this—sort of," but to the point where these techniques can be used easily. You cannot speak a language if you must stop over each word to think about the proper tense or other form; and similarly you cannot do linear programming if you must stop at each step to remember the rules of algebra. You will find most of this chapter and some of the next easy, but each reader must decide how often to pause and to digest thoroughly those passages which are not easy. In particular, when we come to simultaneous equations, we not only introduce a method which may be new to you (and which is of fundamental importance to this text), but we use some sophisticated notational and symbolic devices with which you must be thoroughly familiar if you are to understand the rest of this book. And we introduce the concept of an algorithm, which is utilized throughout the book. You will not be sure that you understand these things until you have done the problems at the end of the chapter (as we said in To the Student, page xi, which we hope you didn't skip).

0.1 ARITHMETIC

The basic operations of arithmetic are addition, subtraction, multiplication, and division. We will be performing these operations over and over again with both integers and fractions (an *integer* is a whole number—positive, negative, or zero—as distinguished from a fraction) and with positive numbers, negative numbers, and zero.

0.1.1 Zero

The introduction of the symbol zero, meaning nothing, into mathematics by the Hindus about 2000 years ago was essential for the development of mathematics. If you don't believe this, try a simple long division using roman numerals, say MCCXLVI divided by XIV. The Romans were forced into this awkwardness because, without zero, one cannot distinguish between, say, 23 and 203. But the use of zero does introduce some complications, which we shall now review.

Arithmetic operations with zero reduce to the following simple rules: (1) When zero is added to a number, the number is not changed; thus, $3+0=3$. (2) When zero is subtracted from a number, the number is not changed; thus $3-0=3$. (3) When a number is multiplied by zero, the product is zero; thus $3\times 0=0$. (4) When zero is divided by any number, the quotient equals zero; thus $0\div 3=0$. (5) Division by zero is not defined and is not permitted.

Special comments are pertinent with respect to rules 3 and 5. A professor of mathematics once made a wager with a nontechnical colleague on the cumulative results of his school's football team for the season. Specifically, he asserted, "Our team is so good that I'm willing to bet five dollars that the sum of their scores during the season will exceed the product of their opponents' scores during the season." The "sucker" did some quick mental arithmetic: "7×7 is 49, 7×49 is 343, 7×343 is a very big number; even if the opponents make only 7 points per game, the product of their scores is bound to exceed any sum that our team could get over the nine games of the season." And he accepted the bet. Of course he had forgotten that regardless of how large the product might become, a single score of zero multiplied into the overall score leaves a final answer of zero. I am happy to report that this particular university did not achieve a shutout against any of its opponents that year, and the mathematics professor lost his wager.

The easiest thing to say about division by zero is that it is forbidden. Later we shall be talking about symbolic division; for example, we shall say that x divided by y is equal to z. The mathematician who makes such an assertion always adds under his breath "provided y is not equal to zero," and if he is cautious, he will invariably make this explicit (by writing $y \neq 0$, where $\neq$ means "is not equal to").

Division is defined by some such statement as the following: "The result of dividing a number x, called the *dividend,* by a number y, called the *divisor,* is a third number z, called the *quotient,* such that if z is multiplied by y it will yield x." If, for example, we let $x = 3$ and $y = 0$, then the result of dividing 3 by 0 must be a number such that when it is multiplied by 0 it yields 3. Clearly there is no such number. We can define a special number called *infinity* which is larger than any ordinary number and say that 3 divided by 0 is equal to infinity. There is a sense in which this is valid. Mathematicians sometimes use the word "infinity" in such context. The trouble is that 4 divided by 0 also equals infinity, and these two infinities are not distinguishable. That is, the product of 0 times infinity is not unique: it may be 3, or 4, or anything else you like, and mathematicians prefer operations which are unambiguous.

0.1.2 Negative Numbers

We will review here without explanation the arithmetic operations with negative numbers which we learned in elementary school.

Addition of a negative number is the same as subtraction of a positive number. Thus $3 + (-2) = 3 - 2 = 1$.

Subtraction of a negative number is the same as addition of a positive number. Thus $3 - (-2) = 3 + 2 = 5$; $-3 - (-2) = -3 + 2 = -1$.

Multiplication of a positive number by a negative number yields a negative product. Thus $(+3) \times (-2) = -6$; $(-3) \times (+2) = -6$.

Multiplication of a negative number by a negative number yields a positive product. Thus $(-3) \times (-2) = +6$.

The rules for division with negative numbers are like the rules for multiplication with negative numbers. Thus $(-6) \div (2) = -3$; $(6) \div (-2) = -3$; $(-6) \div (-2) = +3$.

0.1.3 Fractions

Every real number is either an integer or a fraction. The positive integers are the numbers 1, 2, 3, . . . (note the use in mathematics of three dots to mean "et cetera"). The negative integers are the numbers $-1, -2, -3, \ldots$. Zero is also an integer. Any integer may be expressed in the form of a fraction. Thus $3 = \frac{3}{1} = \frac{6}{2} = \frac{45}{15} = \cdots$. A fraction whose value is greater than unity is called an *improper* fraction, but there is nothing "improper" about it insofar as arithmetic manipulations are concerned. Thus the rules of addition, subtraction, multiplication, and division of fractions are the same whether they are proper or improper.

The value of a fraction is not altered if its numerator and denominator are multiplied by the same number, as indicated in the example $(\frac{3}{1} = \frac{6}{2} = \cdots)$. It is also true that the value of the fraction is not altered if the numerator and denominator are divided by the same number, since division and multiplication are essentially equivalent operations; that is, multiplying numerator and denominator by 2 is the same thing as dividing numerator and denominator by $\frac{1}{2}$. If the numerator and denominator of a fraction are both divisible by the same integer, it is usually desirable to *reduce* the fraction by performing such division. When such division is no longer possible, the fraction is said to have been *reduced to lowest terms*. Thus if the fraction $\frac{6}{14}$ should appear in a computation, it would often be desirable to change it immediately to $\frac{3}{7}$. The reverse operation is usually required when fractions are added; for example if we wish to perform the addition $\frac{3}{7}+\frac{5}{2}$ we cannot immediately add sevenths and halves. However, $\frac{3}{7}$ is equal to $\frac{6}{14}$ and $\frac{5}{2}$ is equal to $\frac{35}{14}$ and we can easily add $\frac{6}{14}$ to $\frac{35}{14}$ to yield $\frac{41}{14}$. The number 14 is called the *common denominator*. A common denominator can always be found by multiplying the denominators of the two fractions to be added, but it is usually convenient to find a smaller common denominator if this is possible, and it will be possible whenever the two denominators have a common *factor;* that is, whenever both denominators can be divided exactly by the same integer. For example, $\frac{5}{6}+\frac{4}{9}=\frac{45}{54}+\frac{24}{54}=\frac{69}{54}$. It will be noted that $\frac{69}{54}$ can be reduced: $\frac{69}{54}=\frac{23}{18}$. The same operation might have been performed more quickly by noting that the *least common denominator* of the two original fractions was 18. Thus $\frac{5}{6}+\frac{4}{9}=\frac{15}{18}+\frac{8}{18}=\frac{23}{18}$.

Subtraction of fractions follows the same set of rules as addition.

To multiply fractions, simply multiply the numerators to yield the numerator of the product, and multiply the denominators to yield the denominator of the product. Symbolically: $(\frac{a}{b})\times(\frac{c}{d})=(a\times c)/(b\times d)$. Thus $\frac{3}{7}\times\frac{5}{2}=\frac{15}{14}$. Note how much more concise the symbolic instruction is than the verbal one.

To divide by a fraction, simply turn the fraction upside down and multiply. Thus $3\div\frac{1}{2}=3\times\frac{2}{1}=6$.

$$\frac{3}{7}\div\frac{5}{2}=\frac{3}{7}\times\frac{2}{5}=\frac{6}{35}$$

0.1.4 Decimal Fractions

Operations with fractions can sometimes rapidly become very messy. For example, $\frac{2}{13}+\frac{3}{17}+\frac{4}{19}=\frac{2271}{4199}$. This fraction has already been reduced to lowest terms, and the next operation on it could easily make it much worse. When this sort of thing happens, it is easier to work with decimal fractions. To convert an ordinary fraction into a decimal fraction, simply perform the indicated division; that is, $\frac{2}{13}=2\div 13$. If you carry out the long division you will get .153846153846.... The division process never terminates, but for most purposes, the first four digits are plenty, and we will write $\frac{2}{13}\approx .1538$, where $\approx$ is read "is approximately equal to." Usually we will not be so precise, and will write $\frac{2}{13}=.1538$ or, better, $\frac{2}{13}=0.1538$, which is less likely to lead to error. Similarly, $\frac{3}{17}=0.1765$ and $\frac{4}{19}=0.2105$. Adding these three decimal fractions yields $0.5408\approx\frac{2271}{4199}$. Each of these numbers has been expressed to four *sig-*

nificant digits. Note that the 0 in front of the decimal point is not considered a significant digit. Nor is any leading zero; thus $\frac{2}{130} = 0.01538$ has been expressed with just four significant digits.

If greater accuracy had been required, we could have carried out the division to more places and carried more than four significant digits. Actually, for most practical work, and for almost all work in this book, three significant digits will in general be sufficient, although it is never harmful and is sometimes helpful to carry four significant digits, especially when the first significant digit is equal to 1, as in the case of $\frac{2}{13} = 0.1538$. The process of shortening the number to fewer significant digits (thus sacrificing some accuracy for more convenience) is called *truncating* if the last digits are merely dropped. This is not usually the proper way. The proper way is *rounding,* in which case the next-to-last digit must be increased if the last digit is greater than 5. Thus 0.1538 becomes 0.154. When the last digit is a 5, the situation is somewhat special. To be absolutely accurate, one should *round down* by truncating the last digit only when the number was a little bit too large. Thus $\frac{3}{17}$ is a little bit less than 0.1765, and this number should be "rounded down" to 0.176, but $\frac{4}{19}$ is a little bit more than 0.2105, and this number should be "rounded up" to 0.211. However, for most purposes it is sufficient to adopt the systematic rule of truncating when the final digit is 5.

Note that three significant digits (four significant digits when the first significant digit is a 1) is about the accuracy that can be achieved with a 10-inch slide rule, and many people used to prefer to do the kinds of arithmetic involved in linear programming problems with the aid of a slide rule. If one is using an electronic calculator, one should usually round to four significant digits when copying the result.

Why three significant digits? Because very rarely in the real world do we have more accurate knowledge of the values of interesting numbers. If you are told that it costs $2.34 to ship 1000 widgets from Chicago to Keokuk, or that the performance rating of Joe Jones on the widget-making machine is 2.34, or that it takes 2.34 pounds of copper to make a widget, you can rarely have much confidence that it isn't 2.33 or 2.35 instead of 2.34—in fact, you're often lucky if it's this close. In such cases, carrying additional digits in the calculation is a waste of time at best and misleading at worst; reporting the answer with so much *precision* implies that you have that much *accuracy.*

The decision as to whether to use common fractions or decimal fractions in a particular problem depends entirely on the problem. If the fractions are simple (that is, if the denominators are small), then the arithmetic is probably easier using common fractions. Thus $\frac{1}{2} + \frac{1}{3} = \frac{5}{6}$ is probably easier than $0.50 + 0.333 = 0.833$. Furthermore, there are complications due to rounding. Thus $\frac{2}{3} = 0.667$; hence $\frac{2}{3} + \frac{2}{3} = 0.667 + 0.667 = 1.334$, whereas the correct answer is $1.333 = \frac{4}{3}$. On the other hand, decimal fractions have the advantage that you can round off to the number of digits that you wish and thus keep the level of complication tolerable. There is no reason why you cannot change back and forth between common and decimal fractions, or use both at once in different parts of the problem, if you want to.

0.2 ALGEBRA

Algebra is, for our purposes, the branch of mathematics in which we use symbols to represent numbers. In high school mathematics, we usually allowed the letters a, b, and c to represent known numbers and the letters x, y, and z to represent unknown numbers. These numbers were then combined in various ways by the usual operations of arithmetic in the form of equations.

We ask you to reread and to meditate upon the preceding paragraph, because in our experience failure to comprehend it is the most serious single difficulty that faces many students using this book. That is, each symbol such as x or a must ultimately represent a number, and most of the mathematics of this book is concerned with assertions that one number equals another number, or that one number is larger than another number, or smaller. When we solve a problem we must begin by defining a set of such symbols, making clear exactly what number each represents (for example, "let x equal the number of widgets manufactured on Tuesday"), as discussed in some detail in Section 4.11. (To be completely accurate, we must add that later we will introduce symbols such as X and A to represent arrays of numbers.)

An *equation* is a statement that two numbers are equal. One of these numbers appears explicitly or symbolically on the left-hand side of an equals sign, and the other appears on the right-hand side of the equals sign. For example,

$7 = 7$	is an equation, although not a very interesting one
$5 + 2 = 7$	is a somewhat more interesting equation
$3x + 1 = 7$	is a still more interesting equation

They all, however, represent exactly the same thing, namely, that seven equals seven. In the last of these equations, x must clearly be equal to two, since

$$3 \times 2 + 1 = 7$$

If x should take on some other value, the equation would no longer be true. Displaying the statement that $x = 2$ is called *solving the equation.*

An equation is solved by deriving other equations from it until the solution is displayed. An equation may be derived from another by adding the same thing to both sides of the equation or by multiplying both sides by the same thing. Thus, we solve $3x + 1 = 7$ by first adding -1 to both sides, thus deriving the new equation $3x = 6$, and then multiplying both sides by $\frac{1}{3}$, thus deriving $x = 2$. Note that we can also add another equation to our first one, since this constitutes "adding the same thing to both sides." For example, if $3x + 1 = 7$ and $2y + 3 = 5$, then we may add these equations to obtain the valid equation $3x + 2y + 4 = 12$.

The advantages of this symbolic notation are that very general statements about a wide variety of different numbers can be summarized in neat notational fashion. For example, although nothing as complicated as quadratic equations will be discussed in this book, you may recall from high school that if we are given the equation

$$ax^2 + bx + c = 0$$

where a, b, and c are known numbers and x is an unknown number, then the solutions to this equation are given by

$$x = \frac{-b \pm \sqrt{b^2 - 4ac}}{2a}$$

That is, if the indicated arithmetic operations are performed on the known numbers a, b, and c, then two numerical values for x will be produced; and if either of these numbers is substituted for x, the original equation will be true, while no other number substituted for x will make the equation true.

A quadratic equation is an example of a *nonlinear* equation, as is any equation which has a square or square root or other complicated function of an unknown number or a product of unknown numbers. In this book we deal only with *linear functions*, things like $6x + 2y$, where the only mathematical operations are multiplying an unknown (or variable) by a known number or adding and subtracting such terms. A *linear equation* has nothing more complicated than linear functions on each side.

In this book we will continue to use the symbols x, y, and z for unknowns, but instead of calling them unknowns, we will in most cases call them *variables*. Similarly, we will continue to use the symbols a, b, and c for most knowns, but instead of calling them knowns, we will in most cases call them *constants*. Furthermore, in high school we almost never were required to deal with more than three unknowns at a time, whereas in linear programming we will frequently be dealing with dozens of variables, and under these circumstances we will quickly run out of letters. For this reason we will often call all the variables in a problem x's, and we will differentiate them by adding to each x a little number written to the right and below, called a *subscript*. Thus the set of variables in a problem will frequently be called $x_1, x_2, x_3, \ldots$. When we wish to refer to one of these x's without specifying which one, we will call it x_j, where the letter j is called an *index*. We may then add the statement $j = 1, \ldots, n$, meaning that there are n different x's. Sometimes we will need more than one subscript to identify a particular x. In this case we will write something like x_{12}, which is read "x sub one two" or just "x one two."

Similarly, we will frequently have a great many constants, all of which might be represented by the letter a (in a given problem) with appropriate subscripts. It will turn out that we will frequently use double-subscript notation for the a's, whereas the letter b will usually have single-subscript notation, and c may have either. Finally we will retain the letters i and j as index symbols. For example, b_i, $i = 1, \ldots, m$, will represent one of the m different b's, and c_j, $j = 1, \ldots, n$, will represent one of the n different c's.

It must be emphasized that this apparent complexity in notation is used not because we want to be complicated but because we want to be simple. We are dealing with very complicated situations, as we shall see; it turns out that it will be essential for us in typical problems to handle dozens of variables and hundreds of different constants, and an organized method of denoting all these numbers is absolutely essential.

0.3 SET THEORY

Much mathematics can be more easily explained through set theory than in other ways, and in recent years even arithmetic has been taught in many elementary schools on the basis of set theory. If you came up through the "new math," you already know more about sets than you need for this book. We do want to define a *set* as a collection of elements which have any particular property or properties of interest to us. For example, the set of all integers greater than 2 and less than 6 consists of the numbers 3, 4, and 5. A set may be *null* (or *empty*) if it has no elements at all. For example, the set of integers which are both greater than 6 and less than 2 is a null set. A set may easily have an infinite number of elements—for example, the set of all integers greater than 2, or the set of all numbers greater than 2 and less than 6.

We shall be particularly interested in this book in nonnegative sets; that is, sets whose elements are either positive or zero (not negative). Notice that the mathematician is careful in distinguishing the words "positive" and "nonnegative." The set of nonnegative numbers and the set of positive numbers both have an infinite number of elements (or members). The nonnegative set has one more member—namely, the number zero.

0.3.1 Inequalities

Sets are often defined by inequalities. For example, the set of nonnegative x's is represented by $x \geq 0$ (read: "x greater than or equal to zero"), while the set of positive x's is represented by $x > 0$ ("x greater than zero"). The same sets could also be defined, respectively, by $0 \leq x$ ("less than or equal to") and $0 < x$ ("less than").

The rules for manipulating inequalities are similar to those for equations, but there are some differences, of which one will be important to us. As with equations, we may add (or subtract) a constant on both sides without affecting the inequality. Thus if

$$3x_1 + 2x_2 \leq 7x_3$$

then it is also true that

$$3x_1 + 2x_2 - 2 \leq 7x_3 - 2$$

Similarly, we may multiply (or divide) both sides by a positive constant; multiplying the first inequality by 2 yields

$$6x_1 + 4x_2 \leq 14x_3$$

But if we multiply or divide by a negative number, then we must reverse the sign of the inequality:

$$-3x_1 - 2x_2 \geq -7x_3$$

If you stop to think about these rules, they are all obvious. Thus, $6 > 4$ also implies $8 > 6$ (adding 2 to each side) and $12 > 8$ (multiplying each side by 2). But if we multiply each side by -1, we get $-6 < -4$, which of course is true: Negative six is a smaller number than negative four. All these rules apply whether the inequality is of the form $<$, $>$, $\leq$, or $\geq$.

0.4 GRAPHS

We could have called this section "Analytic Geometry," but that might have frightened you, because it is the name of a college mathematics course. Nonetheless, it is the subject of this section. Graphs are a kind of geometry, and describing them by equations and other algebraic forms is "analytic." What makes this complicated in college courses is the attempt to graph nonlinear functions. Here we are interested only in the geometric analogs (that is, graphs) of linear functions or, equivalently, in the algebraic analogs of straight lines and planes (a *plane* is a flat two-dimensional surface such as a tabletop, a wall, or a page of this book).

Corresponding to each of our equations there is a graph. For example, Figure 0-1 shows the graph of the equation $2x_1 + 3x_2 = 6$. Note that we have used the symbols x_1 and x_2, as we will typically do throughout this book, rather than the symbols x and y, as we normally did in high school. The horizontal *axis* of the graph is called the *x_1-axis,* and distances measured along (or parallel to) this axis correspond to values of x_1. Similarly the vertical axis is called the *x_2-axis,* and distances measured along this axis correspond to values of x_2. Note therefore that any point on the x_1-axis corresponds to $x_2 = 0$, and that any point on the x_2-axis corresponds to $x_1 = 0$. They intersect at the point $x_1 = 0$, $x_2 = 0$, which is called the *origin.*

On this graph the point P_1 corresponds to $x_1 = 0$, $x_2 = 2$. Note that this point satisfies the equation, since

$$2 \times 0 + 3 \times 2 = 6$$

Similarly the point P_3 corresponds to $x_1 = 3$, $x_2 = 0$, which also satisfies the equation. The point P_2 corresponds to $x_1 = \frac{3}{2}$, $x_2 = 1$, which also satisfies the equation. And the point P_4 corresponds to $x_1 = 4$, $x_2 = -\frac{2}{3}$, which also satisfies the equation.

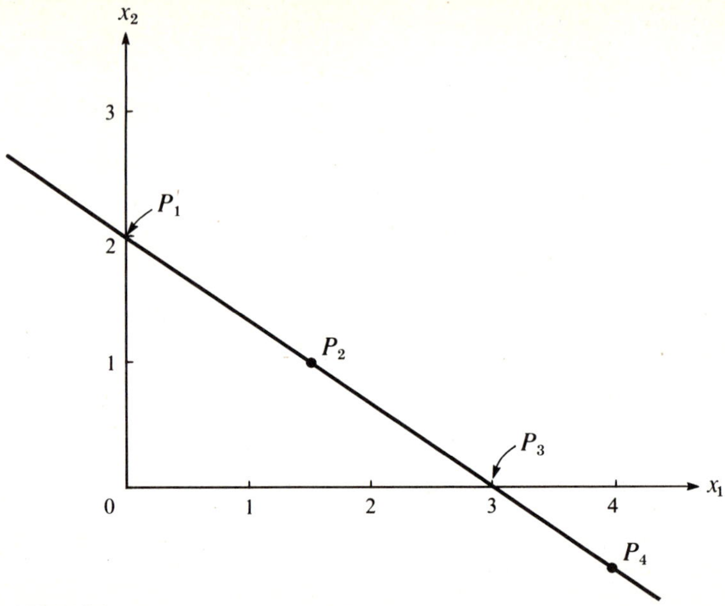

FIGURE 0-1.
Graph of the line $2x_1 + 3x_2 = 6$.

Thus every point on the line represents a pair of values for x_1 and x_2 which satisfies the equation, and every pair of values for x_1 and x_2 which satisfies the equation corresponds to a point on this line. Hence the line corresponds to the equation.

You should be able to interpret and read graphs easily with respect to straight lines. For example, any equation of the form $x_1 = c$ (c is a constant) is represented by a vertical line on the graph. Thus $x_1 = 3$ is a vertical line passing through the point $x_1 = 3$ on the x_1-axis; similarly any horizontal line is represented by an equation of the form $x_2 = c$, where c is some constant (which may be integer or fraction, positive, negative, or zero).

Every such equation (and the corresponding graph) is characterized by a *slope,* which is a measure of the direction of the line. By definition, the slope of a line is the number of units of increase in x_2 for each unit of increase in x_1. Note that the slope of the line in Figure 0-1 is negative since, as x_1 increases, x_2 decreases. Specifically, the slope of this line is $-\frac{2}{3}$.

Throughout this book we will be dealing only with *linear* equations, and the graphs of such equations are uncurved. If there are only two variables in the equation, then the graph of the equation is a straight line.

If we have to graph an equation in three variables, we require three dimensions, that is, ordinary space, as distinguished from the two-dimensional space which appears on a piece of paper. In this case, the x_1-axis goes horizontally to the right as before, the x_2-axis goes vertically upward as before, and now the x_3-axis will come out from the paper toward the reader. We represent

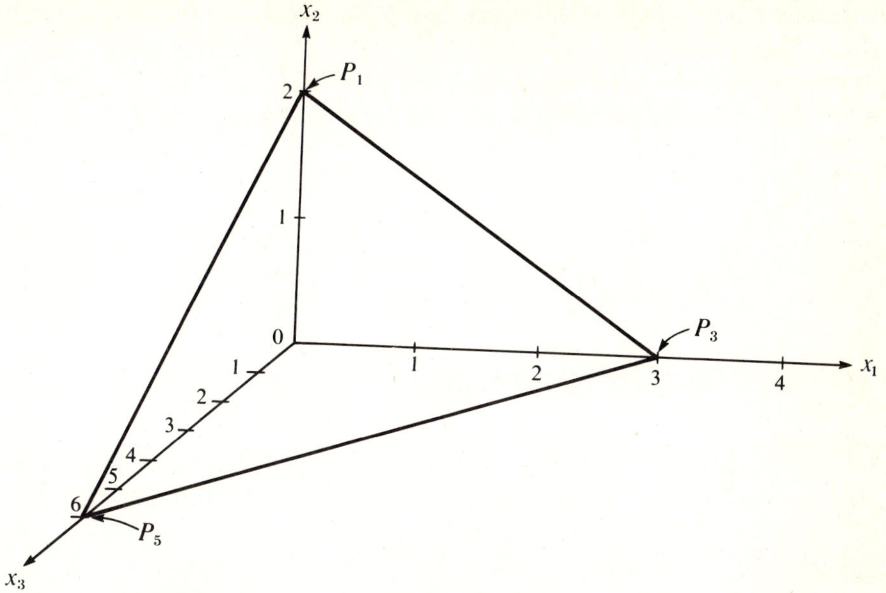

FIGURE 0-2.
Graph of the plane $2x_1 + 3x_2 + x_3 = 6$.

this symbolically by a slanted line, as in Figure 0-2. In this case it becomes confusing if the axes are extended into the negative region. Fortunately, in linear programming we almost never have to consider negative values of any of the variables, and so we may draw the graph shown in Figure 0-2. The equation graphed in this figure is

$$2x_1 + 3x_2 + x_3 = 6$$

In plotting such an equation, we start by letting x_2 and x_3 equal 0, in which case the equation reads $2x_1 + 3 \times 0 + 1 \times 0 = 6$, which leads to the point P_3 where $x_1 = 3$. Similarly we plot P_1 and P_5. Note that when $x_3 = 0$, the line must lie in the plane determined by the x_1- and x_2-axes. In this case, the equation reduces to the previous case, and thus the line connecting P_1 and P_3 on this graph is identical to the same line on the previous graph.

We have shown how, given the equation of a line (or plane), it is possible to graph it by determining the position of two (or three) points. Correspondingly, if we are given the graph of a line (or plane) and are asked to determine its equation, we do this by locating two (or three) points, and again it is easiest if we select those points for which some of the coordinates are 0. For example, suppose we were asked to draw the equation of the line P_1P_3 of Figure 0-1 (or 0-2). We note that the point $x_1 = 3$, $x_2 = 0$ lies on this line, as does the point $x_1 = 0$, $x_2 = 2$. Note that each of these points must fit an equation of the form $x_1 + ax_2 = b$. Plugging into this equation first the point $x_1 = 3$, $x_2 = 0$, we get $3 + 0 = b$. Now plugging in the point $x_1 = 0$, $x_2 = 2$, we get

$0 + 2a = b$. Since we already know that $b = 3$, we now know that $a = \frac{3}{2}$, and therefore we have the equation $x_1 + \frac{3}{2}x_2 = 3$, which is, of course, equivalent to $2x_1 + 3x_2 = 6$. Note in this example that we used the symbols a and b for unknowns while the x's were known. Although certain conventions (such as using x for an unknown) are often practiced, the particular symbols used are arbitrary, and one should not attach too much significance to them.

Unless two planes are parallel, they intersect in the form of a line. In this case, for example, the plane determined by the equation $2x_1 + 3x_2 + x_3 = 6$ and the plane determined by the equation $x_3 = 0$ intersect in the line P_1P_3. Similarly, two lines, if they intersect at all, intersect in a point.

0.5 SIMULTANEOUS EQUATIONS

As we have seen, an equation such as $3x + 1 = 7$ has only a single solution—that is, a single value of x which makes the equation true—whereas an equation such as $2x_1 + 3x_2 = 6$ has many different solutions (in fact, an infinite

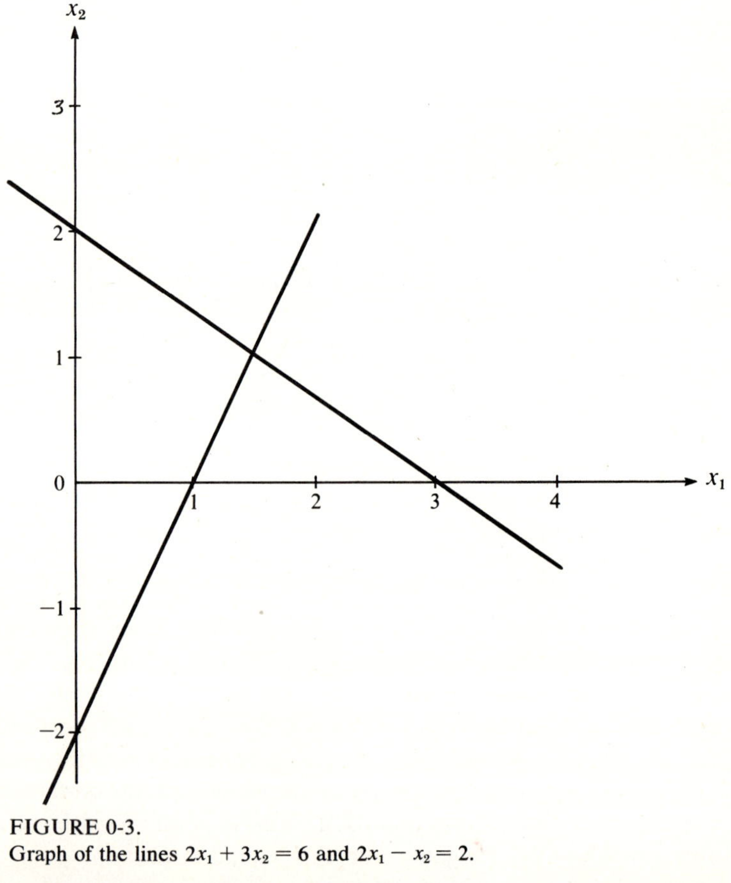

FIGURE 0-3.
Graph of the lines $2x_1 + 3x_2 = 6$ and $2x_1 - x_2 = 2$.

number). Now consider another equation, say, $2x_1 - x_2 = 2$; this equation also has an infinite number of solutions, but most of the pairs of values of x_1 and x_2 which satisfy the first equation do not satisfy the second, and vice versa. If we require that these equations hold *simultaneously,* then only a single pair of values—namely, $x_1 = \frac{3}{2}$, $x_2 = 1$—can be true. The situation is indicated in Figure 0-3, where the two equations are graphed. An infinite number of points lie along each line, corresponding to the solutions to each individual equation, but only a single point lies on both lines. This point represents the solution to the set of *simultaneous equations.* In this case, of course, that solution is $x_1 = \frac{3}{2}$, $x_2 = 1$.

We shall have a great deal more to say about simultaneous equations, but it might not be fair to put it in this chapter, because you may not have seen it all in high school. We need some technical words and symbols, and some precise methods, for describing and manipulating sets of simultaneous equations, and these constitute the principal subject matter of the next chapter.

0.6 IMPORTANT TERMS

Of the terms which are defined in this chapter, the following are most important. Each is in italics where it is first defined. Do not go on to Chapter 1 until you are sure you are thoroughly familiar with all of them.

- Integer
- Infinity
- Reducing (a fraction)
- Common denominator
- Factor
- Significant digits
- Equation
- Linear equation
- Variable
- Constant
- Subscript
- Index
- Set
- Axis
- Origin
- Slope
- Simultaneous (equations)

1

SOME MORE HIGH SCHOOL MATHEMATICS

Detached coefficients The example of two simultaneous equations containing two variables which we introduced at the end of the last chapter is about as simple as you can get with simultaneous equations. In later sections of this book we will be dealing with very complicated sets in which there may be many equations and many variables. In such cases, it saves a great deal of time and space to use a technique known as *detached coefficients*. Consider again the set of equations

$$\begin{aligned} 2x_1 + 3x_2 &= 6 \\ 2x_1 - x_2 &= 2 \end{aligned} \tag{1-1}$$

[Notice the (1-1) on the right-hand side of the page; we shall refer to this set of equations as equations (1-1) or merely as (1-1). It is the first set of formulas in Chapter 1 to which we will subsequently refer.] The numbers 2 and 3 on the left-hand side of the first equation are called the *coefficients* of the variables x_1 and x_2, respectively. Similarly, in the second line the numbers 2 and -1 are the coefficients of the variables x_1 and x_2, respectively. We now "detach" these coefficients and write them without writing the variables:

$$\begin{matrix} 2 & 3 & \quad & 6 \\ 2 & -1 & \quad & 2 \end{matrix} \tag{1-2}$$

It is important that you understand this notational device, since we will be using it many times. The set of numbers still represents precisely the same thing as the preceding set of equations. The first line may be read "two three six," or it may be read "two x-one plus three x-two equals six"; but in either case, the latter is what is meant by the set of numbers.

Matrices Since we will be using such arrays of numbers frequently, we have a special name defined for them: A *matrix* is any rectangular array of numbers. On the left of (1-2) we have a 2×2 matrix; that is, a matrix with two rows and two columns. This particular matrix is *square,* since it has the same number of rows as of columns; however, matrices need not be square. On the right of (1-2) we have a very special kind of matrix consisting of two rows and a single column, and for this we have a special name: A *vector* is a matrix which has only a single row or only a single column; that is, a vector is a row of numbers or a column of numbers. The words "vector" and "matrix" will appear frequently in this book. Any set of numbers written in rows and columns may be considered a matrix; thus, the two rows and three columns of (1-2) may be regarded as a single 2×3 matrix.

A vector is simply a $1 \times n$ or $n \times 1$ matrix. The special case of a 1×1 matrix (which is called a *scalar* by mathematicians, although we will not need to use that word in this book) is merely an ordinary number. There is a branch of mathematics called "matrix algebra" which is very useful in higher mathematics for manipulating matrices. However, no matrix algebra will be used in this book. We are using the words "matrix" and "vector" only as names for arrays of numbers.

Other notational devices are also useful in dealing with matrices. For example, the coefficients of a rectangular matrix will often all be given the symbol a with two subscripts. The first subscript represents the row and the second subscript represents the column. For vectors we will use letters such as b and c with a single subscript. For example, the above set of equations would be notationally represented by

$$\begin{aligned} a_{11}x_1 + a_{12}x_2 &= b_1 \\ a_{21}x_1 + a_{22}x_2 &= b_2 \end{aligned} \qquad (1\text{-}3)$$

If we detach coefficients, this becomes

$$\begin{matrix} a_{11} & a_{12} & & b_1 \\ a_{21} & a_{22} & & b_2 \end{matrix} \qquad (1\text{-}4)$$

Again we stress that this is only a symbolic device; the a's and b's represent some particular numbers. Finally it is often convenient to represent the whole set of equations by an even more compact notation. Specifically, the entire matrix of a's is represented by A, the entire vector of b's by B, and the set of x's (which is itself a vector) by X. In this case our pair of equations may be represented by

$$AX = B \qquad (1\text{-}5)$$

where

$$A = \begin{pmatrix} a_{11} & a_{12} \\ a_{21} & a_{22} \end{pmatrix} \qquad X = \begin{pmatrix} x_1 \\ x_2 \end{pmatrix} \qquad B = \begin{pmatrix} b_1 \\ b_2 \end{pmatrix}$$

This extremely compact description not only represents our set of two equations, but it can represent any set of m equations in n variables. We have gone from the very specific description in (1-1) or (1-2) of a particular set of two equations in two variables to a more general description in (1-3) or (1-4) of any set of two equations in two variables, and finally to a still more general description in (1-5) of any set of m equations in n variables. This will be especially useful to us later when m and n become large numbers.

Finally, we shall occasionally use a different symbol for a matrix, namely, $\|a_{ij}\|$, meaning "a matrix of a_{ij}'s." Then we could write an equation

$$A = \|a_{ij}\|$$

which says nothing except that two symbols are defined to have the same meaning.

Because this row-column notation is unambiguous, it can be utilized to carry a lot of information very compactly. For example, the symbol a_{kk} represents an element in the kth row and kth column; even if we know nothing about k, we know that this element must lie on the diagonal.

Degeneracy Returning to the special case of two equations in two variables: Since each equation is represented on a graph by a line, and since the intersection of two lines is a point, it will in general be true that a set of two simultaneous equations in two variables will have a unique *solution*. However, if the equations are "degenerate" this may not be true. In Figure 1-1 we show a pair of equations having no solution. You can see why this must be true. If the expression on the left-hand side of these equations is equal to 6, it cannot at the same time be equal to 8, and vice versa. The corresponding graph says that the solution of these simultaneous equations is at the point where the two lines meet, and since the two lines are parallel, they never meet. In Figure 1-2 we have a different degenerate situation: The two equations are not really different; in fact, since the second equation can be obtained from the first by multiplying both sides by 2, they are really the same equation written differently. Thus they represent the same line, and any point which satisfies either one of the equations must satisfy the other; there are therefore an infinite number of solutions.

Simultaneous equation sets may have more than two variables, more than two equations, or both. In Figure 1-3 we graph a set of two simultaneous equations in three unknowns. One of these equations is the one which we graphed before in Figure 0-2. Each equation graphs as a plane, and the two planes intersect in a straight line. This line has been dotted and represents the *solution set* of this set of equations, that is, the set of all solutions to both equations. Here,

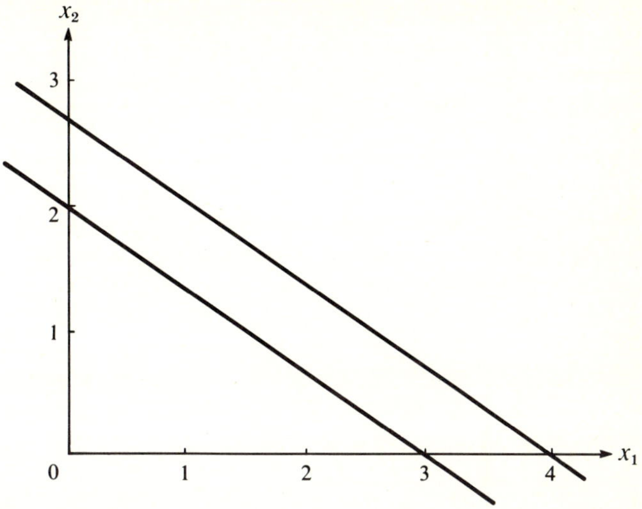

FIGURE 1-1.
Graph of the degenerate set of equations $2x_1 + 3x_2 = 6$ and $2x_1 + 3x_2 = 8$.

of course, there are an infinite number of solutions, since any point on the line satisfies both of the equations. Again, we have drawn only that portion of the space which corresponds to nonnegative values of the variables, since these are the only values in which we will be interested for most of this book.

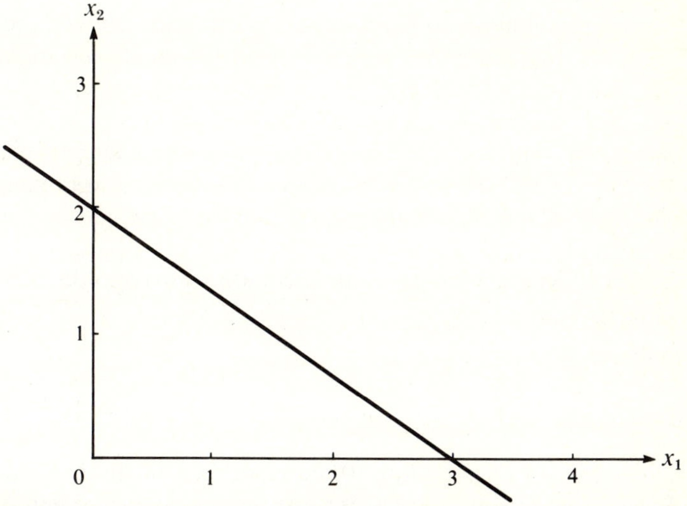

FIGURE 1-2.
Graph of the degenerate set of equations $2x_1 + 3x_2 = 6$ and $4x_1 + 6x_2 = 12$.

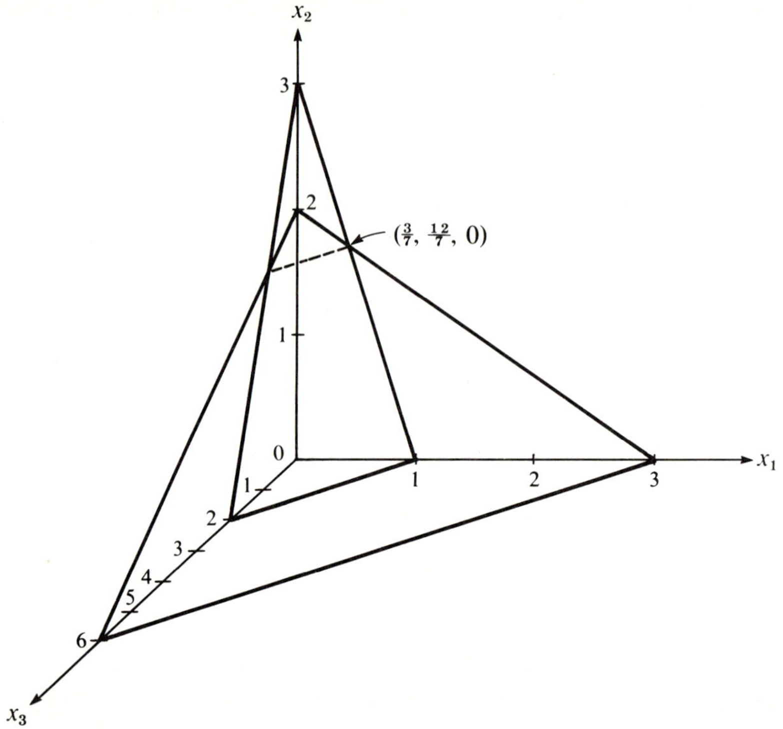

FIGURE 1-3.
Graph of the planes $2x_1 + 3x_2 + x_3 = 6$ and $6x_1 + 2x_2 + 3x_3 = 6$.

You should examine Figure 1-3 until you understand it thoroughly. For example, the plane $2x_1 + 3x_2 + x_3 = 6$ intersects the plane $x_3 = 0$ in the line $2x_1 + 3x_2 = 6$. Similarly, the plane $6x_1 + 2x_2 + 3x_3 = 6$ intersects the plane $x_3 = 0$ in the line $3x_1 + x_2 = 3$. These two lines in turn intersect in the point $x_1 = \frac{3}{7}$, $x_2 = \frac{12}{7}$, $x_3 = 0$, which, if we detach coefficients, may also be written in the form $(\frac{3}{7}, \frac{12}{7}, 0)$, which is a vector. Problem 1-1 asks you to explain the other lines and points on this diagram in the same way; you should try this problem now.

Degeneracy in large sets of equations may not be obvious. For example, consider the set

$$\begin{aligned} 2x_1 + 3x_2 + x_3 &= 6 \\ 6x_1 + 2x_2 + 3x_3 &= 6 \\ 2x_1 + 10x_2 + x_3 &= 18 \end{aligned}$$

This appears to constitute three equations in three unknowns which should have a unique solution; but it turns out that the third equation can be obtained by subtracting the second from four times the first; or the first could be obtained by taking one-quarter of the sum of the last two. Thus, they represent in some sense only two equations: Any two of them represent the same solution set as all three. That is, all three planes intersect in the same line—the dotted line of Figure 1-3—and any two of them define this line adequately. A different type of degeneracy results if we change the 18 in the third equation to some other number. Now each pair of planes intersects in a line; but the three intersection lines are parallel and so have no intersection; hence the three equations are *inconsistent* and have no solution.

We can now define *degeneracy*. The word is used in mathematics to apply to any type of coincidence or unique situation, and we shall meet it repeatedly. But for now, a set of equations is *degenerate* if it is inconsistent or if one or more of the equations can be derived from the others. The reason why we use the same word for both of these conditions is evident from the following set of equations:

$$\begin{aligned} 2x_1 + 3x_2 &= b_1 \\ 4x_1 + 6x_2 &= b_2 \end{aligned}$$

This equation set is degenerate regardless of what numbers b_1 and b_2 are. If $b_2 = 2b_1$, then one equation (either one) is superfluous; if $b_2 \neq 2b_1$, then the set is inconsistent.

Using the notation $AX = B$, we may say that degeneracy is associated with A, not with B. Furthermore, A is degenerate (and so the equation set is degenerate) if and only if a row of A can be derived from other rows. Remember that you may multiply (or divide) any row by any constant you like, and then add it to other rows, to derive a new row. If A is square (n equations in n variables), then the set has a unique solution if and only if A is nondegenerate; if A is degenerate, there is either no solution or an infinite number. Later on we will be dealing mostly with nonsquare matrices, with n columns (variables) and m rows (equations), where $n > m$; but if one row of A can be derived from the others, A is still degenerate and will lead to certain special problems.

Hyperplanes We have noted that a linear equation in two variables graphs as a straight line, and a linear equation in three variables graphs as a plane in three dimensions. It naturally occurs to us that something analogous must represent a linear equation in four variables; however, four-dimensional space is outside our experience (if indeed it exists at all), and there is no way we can graph such an equation in any straightforward way. Nonetheless, it is convenient to talk about higher dimensional spaces and to use them as a conceptual aid in the geometric representation of what is happening in connection with equations representing four or more variables. Thus, we consider four-dimensional space to consist of a space with four mutually perpendicular axes, x_1, x_2, x_3, and x_4. A linear equation in four variables then represents a *hyperplane* in such a space. Two hyperplanes in four dimensions, if they intersect, intersect in a plane; three such hyperplanes intersect in a line; and four such hyperplanes intersect in a point.

In this sense, every set of equations, no matter how complicated, has a geometric analog, and if we happen to have a situation defined by 27 variables we shall consider it to be in 27-dimensional space, whatever that means. You must not think that there is anything complicated about spaces of four or more dimensions. Professional mathematicians cannot visualize them any more than you can; they have learned, however, that it is convenient to think in terms of the geometric representation of situations involving four or more variables and that one can often gain great insight by drawing analogous two-dimensional diagrams. This is what we shall be doing throughout this book.

1.1 SOLUTION OF SIMULTANEOUS EQUATION SETS

We repeat here our first set of simultaneous equations:

$$\begin{aligned} 2x_1 + 3x_2 &= 6 \\ 2x_1 - x_2 &= 2 \end{aligned} \qquad (1\text{-}1)$$

The easiest way to solve this set is simply to subtract the second equation from the first, yielding

$$4x_2 = 4$$

and to divide both sides by 4, yielding

$$x_2 = 1$$

Substituting this back into the first equation we get

$$\begin{aligned} 2x_1 + 3 \times 1 &= 6 \\ 2x_1 \qquad &= 3 \\ x_1 \qquad &= \tfrac{3}{2} \end{aligned}$$

Thus the solution is represented by the point $x_1 = \frac{3}{2}$, $x_2 = 1$, or the vector $(\frac{3}{2},1)$. But it should be noted that $x_1 = \frac{3}{2}$ and $x_2 = 1$ are also the equations of lines, and these lines are graphed in Figure 1-4. Also graphed on the same figure are the original two lines copied from Figure 0-3. It is of great interest to note what we have accomplished here. The solution set of the original pair of (oblique) lines consisted of the point $(\frac{3}{2},1)$. We have manipulated these lines, or equations, by the permissible rules (such as adding one equation to another or multiplying an equation by a constant), and we have achieved a new set of lines which have the same solution set. This new set of lines is more convenient because the coefficient of x_1 is zero in one equation, and the coefficient of x_2 is zero in the other. That is, in place of

$$\begin{matrix} 2 & 3 & & 6 \\ 2 & -1 & & 2 \end{matrix} \qquad (1\text{-}2)$$

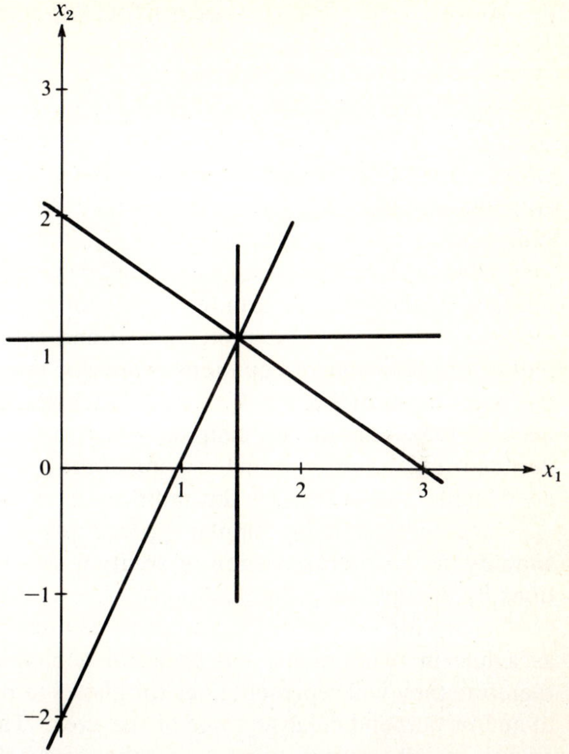

FIGURE 1-4.
The lines of Figure 0-3 and their solution.

we have a new set of equations:

$$\begin{matrix} 1 & 0 & & \frac{3}{2} \\ 0 & 1 & & 1 \end{matrix} \qquad (1\text{-}6)$$

which are perpendicular and parallel to the axes and which represent the same solution set. The matrix of ones and zeros on the left of (1-6) is a sufficiently important special type of matrix to have two names. An *identity matrix* is a square matrix which has 1s everywhere on the northwest-to-southeast diagonal, and 0s everywhere else. On the left of (1-6) we have a 2×2 identity matrix. The 3×3 identity matrix looks like this

$$\begin{matrix} 1 & 0 & 0 \\ 0 & 1 & 0 \\ 0 & 0 & 1 \end{matrix} \qquad (1\text{-}7)$$

If we permute (that is, interchange) some of the rows (or columns) of an identity matrix, we get a *permutation matrix* such as

$$\begin{matrix} 1 & 0 & 0 \\ 0 & 0 & 1 \\ 0 & 1 & 0 \end{matrix}$$

which consists entirely of 1s and 0s with exactly one 1 in each row and in each column. Thus, an identity matrix is a special (and symmetric) form of permutation matrix.

We could do something very similar with the equations of Figure 1-3; by multiplying the first equation by 3 and subtracting the second equation from it, we can obtain the two equations $x_2 = \frac{12}{7}$ and $14x_1 + 7x_3 = 6$. Either the original pair or the new pair of equations represent the identical solution set, namely, the dotted line of Figure 1-3. The latter set, however, is more convenient, because the equations are simpler, which is equivalent to saying that they are perpendicular and/or parallel to some of the axes or that they have several zero coefficients; thus $x_2 = \frac{12}{7}$ is the same as $0x_1 + 1x_2 + 0x_3 = \frac{12}{7}$.

We will be using similar procedures throughout this book, namely, simplifying the representation of solution sets for sets of simultaneous equations by manipulating the equations. These simpler equations will have zero coefficients for several of the variables (as does the equation $x_1 = \frac{3}{2}$, considered as a line in two-dimensional space or a plane in three-dimensional space); therefore they will represent lines (or planes or hyperplanes) which are parallel to and/or perpendicular to some of the axes. The entire set of such simplified equations will continue to represent the same solution set, which will then represent, in linear programming, all the feasible programs of a particular problem. Since we will be doing this with large numbers of equations in large numbers of variables, we will need a more organized procedure, which is the subject of the following section.

1.2 GAUSSIAN ELIMINATION

This method of solving sets of simultaneous equations was developed about 200 years ago by Karl Friedrich Gauss, probably the greatest of all mathematicians. You may have seen it in high school under the name Gauss reduction or Gauss-Jordan method. It is called an *elimination* method because, among other things, it eliminates coefficients (and replaces them with zeros). We will illustrate it by an example, solving a set of three equations in three unknowns, and then write out the rules so that you may use them to manipulate any set of equations.

We will solve the equation set

$$\begin{aligned} 2x_1 + x_2 - 2x_3 &= 10 \\ 3x_1 + 2x_2 + 2x_3 &= 1 \\ 5x_1 + 4x_2 + 3x_3 &= 4 \end{aligned} \tag{1-8}$$

First we detach the coefficients, yielding

$$\begin{array}{ccc|c} 2 & 1 & -2 & 10 \\ 3 & 2 & 2 & 1 \\ 5 & 4 & 3 & 4 \end{array}$$

The objective of this method is to manipulate the A-matrix (the matrix on the left of the vertical line) until it becomes an identity matrix like (1-7). Actually any permutation matrix will do, but it is just as easy and somewhat neater to generate the symmetrical identity matrix. When we have the identity matrix, the equations will have been solved, as we shall see. To commence, we change the first number in the first row into a 1 by dividing the first equation by 2, to get

$$1 \quad \tfrac{1}{2} \quad -1 \qquad 5$$

Remember that this row of numbers represents an equation, namely, the equation $x_1 + \frac{1}{2}x_2 - x_3 = 5$. This row is called the *pivot row,* and the 1 which we have just developed is called the *pivot element* or *pivot.* It is very easy to convert this 1 into anything we want by multiplying it by the appropriate number. For example, the next thing we wish to do is to generate a 0 where the 3 now exists in the second row, first column. In order to do this we multiply our pivot by 3 (and, of course, multiply each other element in the row by 3):

$$(1 \quad \tfrac{1}{2} \quad -1 \qquad 5) \times 3 = (3 \quad \tfrac{3}{2} \quad -3 \qquad 15)$$

We then subtract this from the existing row, which generates a zero where the 3 was. Thus, the new second row is

$$(3 \quad 2 \quad 2 \qquad 1) - (3 \quad \tfrac{3}{2} \quad -3 \qquad 15) = (0 \quad \tfrac{1}{2} \quad 5 \qquad -14)$$

Similarly, we use the new pivot row and the old third row to develop the third row of our new matrix:

$$(1 \quad \tfrac{1}{2} \quad -1 \qquad 5) \times 5 = (5 \quad \tfrac{5}{2} \quad -5 \qquad 25)$$

$$(5 \quad 4 \quad 3 \qquad 4) - (5 \quad \tfrac{5}{2} \quad -5 \qquad 25) = (0 \quad \tfrac{3}{2} \quad 8 \qquad -21)$$

Our second set of equations is now complete:

$$\begin{array}{cccc} 1 & \tfrac{1}{2} & -1 & 5 \\ 0 & \tfrac{1}{2} & 5 & -14 \\ 0 & \tfrac{3}{2} & 8 & -21 \end{array}$$

This second set of equations now becomes the "old" matrix; it contains the "old" rows, and we will develop a "new" third matrix. Having manipulated the first column into the form we wish—namely, with a 1 in the first row and 0s in the other rows—we now proceed to work in the second column. For the second column we wish to develop a 1 in the second row and 0s in the other rows; we therefore begin by working in the second row of the old matrix. To develop the 1 we divide this entire row by $\frac{1}{2}$—or equivalently, multiply it by 2:

$$0 \quad 1 \quad 10 \qquad -28 \qquad \text{(new second row)}$$

This is our new pivot row, and the 1 in the second column is the new pivot element which will be used to help generate 0s elsewhere in the second column. Thus, this pivot row will be multiplied by $\frac{1}{2}$ and subtracted from the old first row to develop the new first row:

$$1 \quad 0 \quad -6 \qquad 19 \qquad \text{(new first row)}$$

Similarly, the pivot row will be multiplied by $\frac{3}{2}$ and subtracted from the old third row to develop the new third row:

$$0 \quad 0 \quad -7 \qquad 21 \qquad \text{(new third row)}$$

Note that, in every case, the pivot row is multiplied by the appropriate amount and then subtracted from the old row; subtraction is never done in the reverse order. The clever part about this is that the 0s and 1s which have already been developed are never lost, since we are always subtracting 0s from them. Note that our third set of equations, which follows, not only has the 1s and 0s in the desired places in the second column, but retains the 1s and 0s which we had previously developed in the desired places in the first column:

$$\begin{array}{cccc} 1 & 0 & -6 & 19 \\ 0 & 1 & 10 & -28 \\ 0 & 0 & -7 & 21 \end{array} \qquad \text{(third set of equations)}$$

Finally, we must work on the third column. We start by developing our pivot in the third row; this is accomplished by dividing the third row by -7, yielding

$$0 \quad 0 \quad 1 \qquad -3 \qquad \text{(new third row)}$$

This new pivot row must first be multiplied by -6 and subtracted from the first row (or—equivalently—multiplied by $+6$ and added to the first row, if you find that easier) to yield

$$1 \quad 0 \quad 0 \qquad 1 \qquad \text{(new first row)}$$

and then the pivot row must be multiplied by 10 and subtracted from the second row to yield

$$0 \quad 1 \quad 0 \qquad 2 \qquad \text{(new second row)}$$

Our final set of equations is, then,

$$\begin{array}{cccc} 1 & 0 & 0 & 1 \\ 0 & 1 & 0 & 2 \\ 0 & 0 & 1 & -3 \end{array} \qquad \text{(fourth set of equations)}$$

Note that we have developed an identity matrix on the left. The set of equations now reads

$$\begin{aligned} 1x_1 + 0x_2 + 0x_3 &= 1 \\ 0x_1 + 1x_2 + 0x_3 &= 2 \\ 0x_1 + 0x_2 + 1x_3 &= -3 \end{aligned}$$

or

$$\begin{aligned} x_1 &= 1 \\ x_2 &= 2 \\ x_3 &= -3 \end{aligned}$$

Obviously, the equations have been solved; the solution is the vector (1, 2, −3).

After solving a problem, it is always desirable to *check* the solution—that is, to demonstrate that it is correct. Thus, if we substitute (1, 2, −3) into $2x_1 + x_2 - 2x_3 = 10$, we get $2 \times 1 + 1 \times 2 - 2 \times (-3) = 10$, which is valid. Similarly, the second and third equations of (1-8) reduce to $1 = 1$ and $4 = 4$, completing the check.

You should make sure that you understand each of the manipulations above before you read on to the following generalized set of rules. Note the general idea: You wish to develop on the left an identity matrix, namely, a square matrix for which $a_{ii} = 1$ and $a_{ij} = 0$ for $i \neq j$. Please meditate on the compactness and unambiguousness of this statement, which utilizes some of the notational tricks that we have defined. Note how much more complicated and much less clear it becomes if we attempt to do the same thing in ordinary words: "We will manipulate the equations by the various permissible operations in such a way as to make the coefficient of each variable either 0 or 1; and, in particular, so that the coefficient of the variable in a particular row will be 1 if the number of the row and the number of the column are the same, and the coefficient of each variable will be 0 in every place where the number of the row is different from the number of the column." We will continue to use our notational devices for three reasons: First, because they are clear, concise, and unambiguous; second, because they are essential for the development of what we shall call algorithms, a subject on which we shall spend much time; and third, because they will be extremely helpful if you are ever required to write computer programs.

The method of developing the identity matrix is as follows: First, we work in the first column; in this column we first generate a 1 in the first row, and then we generate a 0 in each other row. This gives us our second set of equations. We then work in the second column, first generating a 1 in the second row and then 0s in all other rows; this gives us our third set of equations. Then, we work in the third column. And so forth.

1.3 ALGORITHMS

The formal set of rules as they will be written down are in "algorithmic form." The word *algorithm* (like many other words beginning with al) comes from Arabic; it means: A set of rules applied repetitively to solve a problem. For example, you learned in grammar school an algorithm for performing long division. In this algorithm you first performed some preliminary work, such as writing the dividend and divisor in a particular way, drawing certain lines, and adjusting the decimal point. You then took a trial quotient, multiplied it by the divisor, wrote that trial product below the dividend, subtracted, and, if the trial quotient had been correct, carried down the next digit from the dividend, and proceeded to repeat the whole procedure. You are asked in Problem 1-2 to write out precisely this algorithm for long division.

To give you the feeling for an algorithm, we copy the following recipe for fluffy white frosting from the Betty Crocker Cookbook:*

"Mix thoroughly in saucepan: 1 cup sugar, $\frac{1}{3}$ cup water, $\frac{1}{3}$ tsp cream of tartar. Boil slowly without stirring until syrup spins a 6-8 inch thread. Keep saucepan covered for the first three minutes to prevent crystals from forming on the sides of the pan. While syrup is cooling, beat until stiff enough to hold a point 2 egg whites ($\frac{1}{3}$ cup). Pour hot syrup very slowly in a thin stream into stiffly beaten egg whites, beating constantly. Add $1\frac{1}{2}$ tsp vanilla. Beat until frosting holds its shape. Spread between layers and on top and sides." We now translate into algorithmic form:

Step 1 Mix in a saucepan 1 cup sugar, $\frac{1}{3}$ cup water, $\frac{1}{3}$ tsp cream of tartar. Heat covered saucepan until boiling begins. Reduce heat so solution boils slowly. Wait 3 minutes. Uncover. Go to step 2.
Step 2 Perform thread test. If thread is less than 6 inches, go to step 3; if more than 6 inches, go to step 4.
Step 3 Wait 10 seconds. Go to step 2.
Step 4 Turn off heat. Place two egg whites ($\frac{1}{3}$ cup) in bowl. Beat 30 seconds. Go to step 5.
Step 5 Beat whites 10 seconds. Lift beater slowly. If whites stand up to a point $\frac{3}{4}$ inch or higher, go to step 6; if not, go to step 5.
Step 6 Examine syrup. If it is too hot, go to step 7; if it is cool enough, go to step 8.
Step 7 Wait 30 seconds. Go to step 6.
Step 8 Pour hot syrup . . . .

Now, as with any new technique, it seems awkward the first time you try it. The first time you tried to hold a cue stick, or golf club, or violin bow properly, or swim the crawl, or type with all fingers without looking at the keys, it seemed terrible; but you cannot become proficient unless you use proper technique.

The principal advantage of the algorithmic form is that it is unambiguous. (We apologize for the "too hot" and "cool enough" in step 6. Knowing nothing about cooking, we could not translate more precisely. At least the algorithmic form makes our ignorance obvious.) We specified how much longer you must wait (in steps 3 and 7) before trying the test again; in fact, you cannot write the algorithm without either doing this or making the vagueness explicit ("wait a while, then go to step 6"). Note also that we omit-

*Courtesy "Betty Crocker's Picture Cook Book," copyright 1950, General Mills, Inc.

ted "without stirring." In algorithmic form, if we had wanted you to stir, we'd have said so. The recipe form may seem cautious, but it isn't, because it stimulates the following type of uncertainty: "My mother always tapped her foot when she made frosting; the recipe doesn't say not to, so maybe I'd better." But not if we have an algorithm which, by definition, is complete.

Note also the cyclic nature of the algorithm. You will undoubtedly repeat steps 2 and 3 several times. Thus, we would not recommend the algorithm as an actual aid to cooking—you would have to read the instructions too often. This does not affect the usefulness of algorithms for our purpose.

Note finally how every choice is made to depend on a very simple decision. This may be carried to extremes. For example, the writer of an algorithm might not trust the user to carry out an instruction such as "repeat step 3 ten times." Rather, at the end of step 2 the instructions would read something like the following (we have represented the substantive material by dots):

Step 2 .... Set $k = 0$. Go to step 3.
Step 3 Do.... Increase k by 1. If $k \leq 9$, go to step 3; if $k = 10$, go to step 4.

Note how this reduces the counting of repetitions to elementary choices.

As you may have guessed by now, one of the advantages of this form is that it can be converted into a computer program very easily. In fact, before writing any but the most elementary computer program, the programmer always writes his instructions in such an algorithmic form (or in something equivalent called a *flow diagram*).

We present the algorithm for Gaussian elimination in two forms. The first form is entirely words—the kind of thing you have been accustomed to and probably think is easier. But it includes words like "similarly," which means that it is really vague and requires you to exercise judgment, and therefore it must ultimately be confusing. The second form is precise. After you have mastered it (which may require going through it several times with examples), you will find it better and easier.

1.3.1 Algorithm for Gaussian Elimination, Applied to the Solution of n Simultaneous Equations in n Variables (Vague Form)

Detach the coefficients. Divide every element in the first row by the element in the first column (so that the new element in the first row and first column is 1), and write the quotient as the first row of the second matrix. Now, working on the side, copy the second row, write below it the product of the new first row and the element in the first column of the second row, and subtract. The difference (which has 0 in the first column) is then copied as the second row of the second matrix. Similarly, generate the remaining rows of the second matrix: Copy the next old row on the right; multiply the new first row by the element in the first column of the old row, and write the product below the old row; subtract; copy the difference as the next row of the second matrix. After this second matrix is completed, work in the second column to generate the third matrix. Start by dividing the second row of the second matrix by the element in its second column (so that the new element in the second row and second column is 1), and write the quotient as the second row of the third matrix. Then generate the new first, third, and additional rows of the third matrix, always by copying the old row, subtracting from it the product of the new second row by the element in the second column of the old row so as to generate a 0 in the second column of the new row. After the third matrix is complete, work in the third column to generate the fourth matrix, starting by generating a 1 in the third row, third column, then generating zeros elsewhere in the third column. And so forth.

1.3.2 Algorithm for Gaussian Elimination, Applied to the Solution of n Simultaneous Equations in n Variables (Precise Form)

Step 1 Detach the coefficients. This yields an $n \times n$ matrix (of which the element in the ith row and jth column is called a_{ij}), with an n-row vector to its right (of which the element in the ith row is called b_i). Taken together, these constitute an $n \times (n+1)$ matrix which is called *the old matrix*. Set $k = 0$. Go to step 2.

Step 2 Increase k by 1. Locate a_{kk} in the old matrix. If $a_{kk} = 0$, go to step 8. If $a_{kk} \neq 0$, go to step 3.

Step 3 Divide the entire kth row by a_{kk}, and write the result as the kth row of the new matrix, whose elements are called a'_{ij} and b'_i. (Thus, $a'_{kj} = a_{kj}/a_{kk}$ and $b'_k = b_k/a_{kk}$.) Note that $a'_{kk} = 1$. Set $i = 1$. Go to step 4.

Step 4 If $i = n + 1$, go to step 5. If $i \leq n$, go to step 6.

Step 5 The old matrix is no longer useful. The new matrix is now complete. If $k = n$, stop. If $k < n$, go to step 2, with the just-completed matrix being the old matrix (that is, what was a'_{ij} now becomes a_{ij}).

Step 6 If $i \neq k$, go to step 7. If $i = k$, increase i by 1 and go to step 4.

Step 7 Multiply the kth row of the new matrix by a_{ik} and subtract this product row from the ith row of the old matrix. Write the difference as the ith row of the new matrix. (Thus, $a'_{ij} = a_{ij} - a_{ik} \times a'_{kj}$ and $b'_i = b_i - a_{ik} \times b'_k$.) Note that $a'_{ik} = 0$. Increase i by 1. Go to step 4.

Step 8 Set i equal to k. Go to step 9.

Step 9 If $i = n$, go to step 11. If $i < n$, go to step 10.

Step 10 Add 1 to i. If $a_{ik} = 0$, go to step 9. If $a_{ik} \neq 0$, go to step 12.

Step 11 The equation set is degenerate. Stop.

Step 12 Interchange the ith and kth rows. Go to step 3.

We have warned you that, whenever you divide, you must first ensure that the divisor is not zero. We have done this in step 2, preparatory to dividing by a_{kk} in step 3. If $a_{kk} = 0$, then you must do something about it by means of a subsidiary algorithm called a *subroutine* before going to step 3. Usually this subroutine will not be needed, but for the sake of completeness we present it as steps 8 through 12.

Note that k generally is the column number (and i the row number) in which we are working, but it is convenient in steps 3 and 7 to refer to "the kth row." Reading this algorithm is confusing until it is tried with an example, especially since the steps are not used exactly in the order in which they are written. Go back through the example, and note that the algorithm proceeds as follows: Step 1, set $k = 0$, go to step 2. Set $k = 1$, go to step 3. Compute first row, set $i = 1$, go to step 4. Go to step 6. Set $i = 2$, go to step 4. Go to step 6. Go to step 7. Compute second row, set $i = 3$, go to step 4. Go to step 6. Go to step 7. Compute third row, set $i = 4$, go to step 4. Go to step 5. This matrix is complete, go to step 2. Set $k = 2$, go to step 3. Compute the second row of a new matrix, set $i = 1$, go to step 4. And so forth.

1.4 EXTRA VARIABLES

In high school, if we had n variables we always hoped we had n equations. Thus, for example, three equations and four variables was no good because the system was *indeterminate;* that is, it would generally have an infinite number of solutions, and we couldn't determine a unique solution. Similarly, three equations and two variables was no good because it would generally be overconstrained and thus had to be inconsistent.

In linear programming, on the other hand, we will always be dealing with situations in which the number of variables exceeds the number of equations. The notation we shall use is: n variables and m equations, and $n > m$. We are interested in such "indeterminate" situations because we have become more sophisticated: We are not interested in simple situations where there is a unique solution; we are interested in more complicated situations where there are a very large number (often an infinite number) of possible solutions, and therefore we have a *decision problem.* We are interested in choosing a particular solution which is better than others in accordance with some criterion or objective. Usually this involves maximizing or minimizing an *objective function,* which is a quantitative measure of our profit or cost.

Note that we talked earlier about *the* solution to a set of simultaneous equations; now we are talking about *a* solution. We can now redefine: A *solution* to a set of equations (or to any problem) is a set of numerical values of all the variables. If we wish to emphasize that it satisfies all the equations, we will call it a "feasible solution." If it also optimizes the objective function, then it becomes "the solution" to the problem, or more formally, "the optimal feasible solution."

If we have one equation and two variables, we can always solve for one of the variables *in terms of* the other. That is, we can manipulate the equation so that it is very easy to determine the value of one variable if we are given the other. For example, consider the first equation which appeared in this chapter:

$$2x_1 + 3x_2 = 6$$

This can be solved for x_1 in terms of x_2 by subtracting $3x_2$ from both sides and then dividing both sides by 2 to yield

$$x_1 = 3 - \tfrac{3}{2}x_2$$

Alternatively, it can be solved for x_2 in terms of x_1 to yield

$$x_2 = 2 - \tfrac{2}{3}x_1$$

Either of these forms may be very convenient.

In the same way, if we have a large set of m equations in n variables, where $n > m$, we can solve for any m of the variables in terms of the remaining $n - m$ variables. Furthermore, this can be done by Gaussian elimination without changing the above algorithm in any way. Suppose, for example, we were given two equations in three variables and that these two equations were the first two equations of the set (1-8), namely,

$$\begin{aligned} 2x_1 + x_2 - 2x_3 &= 10 \\ 3x_1 + 2x_2 + 2x_3 &= 1 \end{aligned}$$

Note in this case that $n = 3$, $m = 2$, and $n - m = 1$. This set of equations has an infinite number of solutions; we can arbitrarily pick any value we like for any variable and then find a solution. Suppose we plan to pick the value of x_3, but we haven't yet decided what it will be. It will be convenient to solve for x_1 and x_2 in terms of x_3 so that when we do pick the value of x_3 it will be easy to determine the solution. This may be done by the Gaussian elimination algorithm, and in fact involves no new arithmetic whatsoever. It will yield precisely the first two rows of our third matrix in the example above. These state:

$$\begin{aligned} 1x_1 + 0x_2 - 6x_3 &= 19 \\ 0x_1 + 1x_2 + 10x_3 &= -28 \end{aligned} \qquad (1\text{-}9)$$

which can also be written in the form

$$\begin{aligned} x_1 &= 19 + 6x_3 \\ x_2 &= -28 - 10x_3 \end{aligned}$$

If we now specify a value for x_3, it will be easy to determine the corresponding values of x_1 and x_2.

A set of m equations in n variables (where $n > m$) which have been solved for m of the variables in terms of the other $n - m$, as in (1-9), is said to be in *canonical form*. An equivalent definition of canonical form is: A matrix of m rows and more than m columns in which m of the columns constitute a permutation matrix. Do you see why these definitions are equivalent? We apologize for the fancy word "canonical"; it just means "standard." But since we shall use the phrase "standard form" to mean something else, we decided to call this the "canonical form," which is what everybody else calls it.

The m variables in the canonical form which have the 1 and 0 coefficients are called *basic variables*. Mathematicians (by now you could guess this) call the other $n - m$ variables *nonbasic variables*. Thus, we could define the canonical form as representing solution for the basic variables in terms of the nonbasic variables.

Now suppose that we have obtained the canonical form (1-9) and then change our minds and decide that we want to solve for x_2 and x_3 in terms of x_1. We could do this by starting with the original problem and going through two sets of matrices, but it is much easier to start with the form we have already achieved—namely, (1-9)—applying the Gaussian elimination algorithm and working in the third column. We shall have to make a choice where in the third column we want our pivot, but this is obvious: We will want it where the one now is in the first column, namely, in the first row.

We have developed an important principle: If we have a canonical form, and we wish to introduce one new variable into the set of basic variables (for example, above we wish to introduce x_3, which had been nonbasic, and allow x_1, which had been a basic variable, to become nonbasic), then we need to go through the Gaussian elimination only once in order to develop the representation in terms of the new permutation matrix. We need to modify the algorithm only to the extent of identifying the pivot row; the pivot in the incoming basic column (the variable that was nonbasic and will become basic) is in the same row as the 1 in the outgoing basic column (the variable that was basic and will become nonbasic). This new pivot element will usually not be on the diagonal, so we cannot call it a_{kk}; the exact notation will be developed later.

Suppose, for example, that we had manipulated 10 equations in 17 variables to solve for the variables x_1 through x_{10} in terms of variables x_{11} through x_{17}, with the idea in mind that x_{11} through x_{17} would eventually be chosen in some way; that is, we had a canonical form with x_1 through x_{10} basic and x_{11} through x_{17} nonbasic. Now we might change our minds and decide that we wanted instead to choose eventually x_2 and the variables x_{11} through x_{16}; that is, we want x_{17} to become basic and x_2 to become nonbasic. This is done by one application of the Gaussian elimination algorithm, working in column 17 with the second row as the pivot row. Any such application of the Gaussian elimination algorithm to a matrix to make one new column basic is called an *iteration*, and the variable which becomes basic is called the *incoming variable*. We will work this out in detail in Chapter 4.

There is one more term which we must define concerning such sets of equations. We have stated that we may arbitrarily choose the values of $n - m$ of the variables; it is frequently desirable to choose them as 0. In particular, if we have developed a canonical form and set all the nonbasic variables equal to zero, the total solution is completely displayed. Any solution in which $n - m$ of the variables are equal to zero is called a *basic solution.* (It may be that, when we solve for the remaining m variables, some of them are also equal to 0; this is a special kind of degeneracy which we shall meet again.) Basic variables and basic solutions go beautifully together; that is, if we set all the nonbasic variables equal to zero, then we have a very particular basic solution which is completely displayed. Consider (1-9). There are three basic solutions to this set of equations, but the one of particular interest is the one in which $x_3 = 0$; in that case we can immediately read off the solution: $x_1 = 19$, $x_2 = -28$. Much of the arithmetic manipulation involved in linear programming is just that: We will find a basic solution by developing a canonical form by Gaussian elimination, and then we will change one nonbasic variable to basic and go through one more iteration in order to find a new canonical form and thus a new basic solution.

1.5 SUMMATION NOTATION

In this book we will need only one more piece of symbolism, which you have probably seen before, but with which you must be thoroughly familiar from here on. This is the use of the Greek capital letter sigma (Σ) to represent summation. For example, we will write

$$\sum_{i=1}^{3} x_i$$

which is read "summation from i equals one to three of x_i" and which means $x_1 + x_2 + x_3$. More specifically, Σ means: Whatever follows this sign should be written a number of times (as further indicated), with plus signs between. Whenever you see the symbol i (or whatever the index under Σ is), change it to a number. The first time, change it to 1 (or whatever the number following $i =$); the next time change it to the next highest integer; and continue in this way until you have written the term with i replaced by 3 (or whatever the number above Σ is).

Sometimes we will abbreviate this symbolism and simply write

$$\sum_i x_i \qquad \text{or} \qquad \sum x_i$$

We may use this abbreviated form either for convenience, when what is left out is obvious, or because we are purposefully being somewhat vague. For example, we might be a little unsure ourselves as to what the upper limit is, in which case

$$\sum_i x_i = \sum x_i = x_1 + x_2 + \cdots$$

On the other hand

$$\sum_i x_{ij} \qquad \text{and} \qquad \sum_j x_{ij}$$

are very different. The first sum means $x_{1j} + x_{2j} + \cdots$ and represents, for example, the sum of all the numbers in the jth column of the matrix. For example, if $j = 3$ this would represent $x_{13} + x_{23} + \cdots$ and would represent the sum of the third column of the matrix. The second sum above represents $x_{i1} + x_{i2} + \cdots$ and represents the sum of the numbers in the ith row of the matrix.

Now see if you can recognize something more complicated, such as the following

$$\sum_i x_{ij} = 1 \qquad j = 1, \ldots, n$$

It is n different equations, the first of which says

$$\sum_i x_{i1} = 1$$

which means $x_{11} + x_{21} + \cdots = 1$; that is, if you add up all the x's in the first column, the sum is equal to 1. The second equation says that if you add up all the x's in the second column the sum is equal to 1. There is an equation like this for each column up to n. These equations hold, incidentally, for any permutation matrix.

Try another one:

$$\sum_{i=1}^{n} \sum_{j=1}^{n} c_{ij}x_{ij} \qquad \text{or} \qquad \sum\sum c_{ij}x_{ij}$$

You start at the left and set $i = 1$; you go on and set $j = 1$. That means that c_{ij} is c_{11} and it must be multiplied by x_{11}. Now since we have nothing else to do we go back toward the left until we hit the summation sign and we change j to 2; that means that we multiply c_{12} by x_{12} and add this product to the first one. We continue to do this until we have the sum of n products, all of which have $i = 1$ in their subscripts. Now we go back to the outside summation sign, change i to 2, and start all over again with $j = 1$; the net result of all this will be that we will have the sum of sums of all such products. That is, if we had taken each c_{ij} in the c matrix and each x_{ij} in the x matrix and multiplied them together so that we had a $c_{ij}x_{ij}$ matrix, then this double sum would represent the grand summation of all the elements in that matrix.

1.6 IMPORTANT TERMS

Of the terms which are defined in this chapter, the following are most important. Each is in italics where it is first defined. Do not go on to Chapter 2 until you are sure you are thoroughly familiar with all of them.

- Coefficient
- Detached coefficients
- Matrix
- Vector
- Solution
- Solution set
- Degeneracy
- Permutation matrix
- Gaussian elimination
- Pivot
- Check (a solution)
- Algorithm
- Canonical form
- Basic variable
- Nonbasic variable
- Basic solution
- Iteration
- Incoming variable

Problems

Before trying the problems, we suggest that you go back and reread To the Student, page xi. And we remind you again that most solutions are incomplete unless they include a check. Some problems, such as the first one below, do not lend themselves to a formal checking process; but Problem 1-3, for example, must be checked.

1-1 Write the equations for the lines (other than the dotted line and the axes) in Figure 1-3 and their intersections, if any.

1-2 Write out precisely the algorithm for long division as you learned it in high school. Try your algorithm on 748238.0 divided by 34000

You will find this problem incredibly difficult, because long division is really very complicated; it only seems simple to you because you are familiar with it. You must take care of things like the decimal point, the possibility that your trial quotient may be too small or too large, the possibility that the process may come to an end at any time (exact divisibility) or go on indefinitely, and so forth. Our version of this algorithm, which appears in your instructor's manual, required 19 steps, of which the first seven merely adjusted the decimal point (because, for example, something like $260.0\overline{)0.372}$ must be changed to $26\overline{)0.0372}$ before you go any farther). We repeat here step 13 to give you the flavor of it: "Multiply the trial quotient by the divisor, writing this 'trial product' immediately below the trial dividend, with its rightmost digit directly below the trial quotient. If the trial product is smaller than the trial dividend, go to step 15. If it is larger, go to step 14. If they are equal, go to step 18." Here the terms "trial dividend" and "trial product" have been used for the successive lines written at the bottom of the problem and "trial quotient" for the most recent digit in the quotient. Note the precise use of the terms "immediately below" to indicate the next row and "directly below" to indicate the same column. We really believe that after you try this problem, and then compare your answer with ours, you will be convinced that Gaussian elimination is easier than long division.

1-3 Solve the following equation set by Gaussian elimination:

$$\begin{aligned} 2x_1 + 3x_2 &= 5 \\ 4x_1 + 3x_2 &= 4 \end{aligned}$$

1-4 Solve the following system of equations (another name for an equation set) by Gaussian elimination:

$$\begin{aligned} x_1 + 2x_2 + 3x_3 &= 5 \\ 3x_1 + 2x_2 + x_3 &= 15 \\ x_1 - x_2 + x_3 &= -3 \end{aligned}$$

The first operation will involve dividing the first row by one—that is, copying it.

1-5 (*a*) Try the Gaussian elimination method on the following set of equations:

$$\begin{aligned} x_1 + 2x_2 - 3x_3 &= -1 \\ 3x_1 - x_2 + 2x_3 &= 7 \\ 5x_1 + 3x_2 - 4x_3 &= 5 \end{aligned}$$

Actually, this set is degenerate. How does the Gaussian elimination method show this?

(*b*) What do you think would happen if b_3 in this set were 4 instead of 5? Try it.

1-6 (*a*) Write out in detail the formulas corresponding to the following sums

(i) $\sum_{i=1}^{3} a_i$ (ii) $\sum_{k=1}^{3} a_{ik}$ (iii) $\sum_{i=1}^{3} a_k$

(iv) $\sum_{i=1}^{3} (a_i + 7)$ (v) $\sum_{i=1}^{3} a_i + 7$ (vi) $\sum_{i=1}^{3} 7$

(*b*) Using summation notation, write the most compact formula possible for $a_1 + a_2 + a_3 + b_1 + b_2 + b_3 + b_4 + c_{11} + c_{12} + c_{13} + c_{14} + c_{21} + c_{22} + c_{23} + c_{24} + c_{31} + c_{32} + c_{33} + c_{34}$.

Drill Problems

1-7 Write an algorithm for shaving or for applying cosmetics, whichever you are more familiar with. Shaving should be wet and include such things as checking blade sharpness. Cosmetics might concentrate on eye make-up and include provisions for correcting errors.

1-8 (*a*) Solve and check

$$\begin{aligned} x_1 + 2x_2 + 3x_3 &= 4 \\ 2x_1 + 3x_2 + 4x_3 &= 5 \\ 3x_1 + 7x_2 + 5x_3 &= 8 \end{aligned}$$

(*b*) Graph these equations (in the nonnegative region only). It is suggested that you use several different colors of pencil and/or ink—say, one for the axes and one for each of the three planes which correspond to the three equations above. As in Figure 1-3, use solid lines for intersections of these planes with the planes $x_1 = 0$, $x_2 = 0$, and $x_3 = 0$, and use dotted lines for the intersections of the three planes with one another. Note that the dotted lines do not intersect in the nonnegative region (why?).

(*c*) Write the equations for each solid line in your diagram, and solve for the points where they intersect.

1-9 (*a*) Solve the following equations for x_1 and x_2 in terms of x_3 and x_4 by Gaussian elimination:

$$\begin{aligned} 2x_1 + 4x_2 + 6x_3 + 2x_4 &= 36 \\ 3x_1 + 9x_2 + 3x_3 + 6x_4 &= 48 \end{aligned}$$

(*b*) Using only one further iteration, solve for x_1 and x_3 in terms of x_2 and x_4.

Some Extra Drill Problems

(*a*)
$$\begin{aligned} 2x_1 + x_2 - 2x_3 &= 10 \\ 3x_1 + 2x_2 + 2x_3 &= 1 \\ 5x_1 + 4x_2 + 3x_3 &= 4 \end{aligned}$$

Solution: $(1, 2, -3)$

(*b*)
$$\begin{aligned} x_1 + 2x_2 - 3x_3 &= 6 \\ 2x_1 - x_2 + 4x_3 &= 2 \\ 4x_1 + 3x_2 - 2x_3 &= 14 \end{aligned}$$

Degenerate

(*c*)
$$\begin{array}{rrrrr} 2 & 1 & 5 & 1 & 5 \\ 1 & 1 & -3 & -4 & -1 \\ 3 & 6 & -2 & 1 & 8 \\ 2 & 2 & 2 & -3 & 2 \end{array}$$

$(2, \frac{1}{5}, 0, \frac{4}{5})$

(*d*)
$$\begin{aligned} x_1 + 2x_2 - 3x_3 &= 0 \\ 2x_1 + 5x_2 + 2x_3 &= 0 \\ 3x_1 - x_2 - 4x_3 &= 0 \end{aligned}$$

(*e*)
$$\begin{aligned} x_1 + 2x_2 - 3x_3 + 2x_4 &= 2 \\ 2x_1 + 5x_2 - 8x_3 + 6x_4 &= 5 \\ 3x_1 + 4x_2 - 5x_3 + 3x_4 &= 7 \end{aligned}$$

You can make up your own problems if you like, but they might involve horrid fractions. To avoid this, choose AX and compute B; that is, write the left-hand sides, choose a set of x's, and multiply out to find the right-hand sides. Then pretend you didn't know the x's and solve for them.

2
THE ASSIGNMENT PROBLEM

The prerequisite for reading most of this chapter is summarized in pages 2 and 3; we will be adding and subtracting small whole numbers not exceeding 9. However, in Section 2.3 we shall summarize (in matrix notation) what we have done, and it will turn out that underlying our labors are some very sophisticated mathematical concepts.

The assignment problem is to determine the best possible assignment of several workers to an equal number of jobs. Specifically, we shall make the following assumptions: There are n workers ($i = 1, \ldots, n$) and n jobs ($j = 1, \ldots, n$). The performance of each worker on each job can be measured quantitatively, and all such numbers have been supplied. Note that such numbers can be given in terms of performance, in which case we would like to pick numbers as large as possible, or in terms of cost, in which case we would like to pick numbers as small as possible. We shall see in Section 2.6 how one of these can be easily converted into the other. Meanwhile, we shall assume that we are given a set of numbers, c_{ij}, each of which represents the cost to us of putting the ith worker on the jth job (note that there are n^2 such numbers). Our problem is to assign workers to jobs in such a way as to minimize the sum of the c_{ij}'s for the chosen assignment.

Worker \ Job	1	2	3
1	2	✓6	9*
2	3	4*	✓8
3	✓5*	7	1

FIGURE 2-1
An assignment problem.

This problem, while comparatively simple, is typical of the problems we shall be studying in this book. It comes under the heading *optimization under constraints*. Optimum is a Latin word meaning "best," and to *optimize* something means to make the best possible choice. Here we wish to pick the cheapest possible assignment. A *constraint* is a rule which tells us what we must (or cannot) do. For example, we are constrained to fill each job but not to assign more than one worker to the same job. We shall see below how these criteria and constraints are formulated quantitatively.

We begin with the assignment problem shown in Figure 2-1. This figure represents the cost of assigning three workers to three different jobs. For example, to assign the second worker to the first job costs three units of money (dollars per day, rubles per week, or whatever). To assign this same worker to jobs two or three would cost four or eight, respectively. However, we cannot simply assign the second worker to the first job, which would be the best place, because that would mean we couldn't assign the first worker to the first job, which might be even better. A typical assignment is shown by the asterisks, or stars, in Figure 2-1, and you can see that the cost of this assignment is $5 + 4 + 9 = 18$. Alternatively, we could use the assignment shown by the check marks, at a total cost of 19, which is not as good.

Figure 2-1, together with the idea of choosing a particular set of elements such as those shown in Figure 2-1, represents a *model* of the decision problem of optimally assigning three workers to three jobs. More specifically, it is a *mathematical model*. The principal subject of this book is mathematical models and their application to certain types of decision problems. Linear programming is a special case of the broad and important technique of mathematical modeling.

In Figure 2-1 we have shown two different assignments. How many different assignments are possible? If we assign the first job first, it can be assigned to any one of three workers. If it is assigned (as we have assigned it) to the third worker, then the second job can be assigned to any one of the two remaining workers, for a total of six different possible assignments. Having chosen workers for the first two jobs, we have only one worker left for the third job, and the total number of different possible assignments is therefore

$1 \times 2 \times 3 = 6$. The mathematician (as usual) has defined a special name for such a product of successive integers—namely, *factorial*—and uses the symbol ! to represent it. Thus, we may write $3! = 1 \times 2 \times 3 = 6$.

Now suppose there had been a fourth worker and a fourth job. In this case the first job could have been assigned to any one of four different workers; the next job to any one of the three remaining workers, for a total thus far of 12 possibilities; the third job to any one of the two remaining workers, for a total of 24 possibilities; and the fourth job to the one remaining worker. Thus, the number of possible assignments is $4! = 24$; and in general, if there were n jobs and n workers, the number of possible assignments is $n!$.

Clearly, our simple problem of Figure 2-1 could have been solved by enumerating the six different assignments, computing the cost of each, and selecting the smallest total cost. However, the factorials go up rapidly: If there had been 10 workers and 10 jobs, then the total number of assignments to be enumerated would have been $10! = 3{,}628{,}800$; and if there had been 25 workers and 25 jobs, then the number of different assignments would have exceeded 10^{25} (that is, one followed by 25 zeros, which is 10 million million million million). Since there have been less than 10^{25} microseconds (millionths of a second) since the creation of the universe (and we refer here to astronomical creation, not biblical creation), it is clear that even a very powerful computer would not be of much help if we wished to solve a 25×25 assignment problem by "brute force" methods of exhaustive enumeration. It behooves us, instead, to examine the mathematical properties of assignment problems; and it turns out that such examination will be rewarded by the development of a clever method which, as we shall see, permits us to solve a 10×10 problem with only a few minutes of hand work or a fraction of a second of computer time.

We start by rewriting the matrix of Figure 2-1 on the left of Figure 2-2, leaving out the numbers of the workers and jobs, which add no useful information. Of course, we can no longer tell whether workers are rows and jobs are columns, or the other way around; but this doesn't matter. There are no differences, mathematically, between the rows and the columns of an assignment problem. This interchangeability of rows and columns is a most important property of many of the phenomena we shall study in this book and is given the name *duality*. We shall refer to duality frequently.

We present in Figure 2-2 the first assignment we tried, whose total cost is 18. Immediately to the right of this we write a different matrix, whose first two rows are identical, but whose third row differs because each element has been reduced by one; this is indicated by the -1 on the arrow. Of course the relative costs in the reduced third row have not been changed. Note that the starred assignment in the second matrix costs 17, one less than the cost of the corresponding assignment in the original matrix. In fact, this must be true of any assignment: Since any assignment must use exactly one element from the third row, whatever element is chosen must cost one less in the second matrix than in the first, and therefore the total assignment cost must be one less in the sec-

$$\begin{matrix} 2 & 6 & 9^* \\ 3 & 4^* & 8 \\ 5^* & 7 & 1 \end{matrix} \xrightarrow{-1} \begin{matrix} 2 & 6 & 9^* \\ 3 & 4^* & 8 \\ 4^* & 6 & 0 \end{matrix} \xrightarrow{-5} \begin{matrix} 0 & 4 & 7 \\ 0 & 1 & 5 \\ 4 & 6 & 0 \end{matrix} \xrightarrow{-1} \begin{matrix} 0^* & 3 & 7 \\ 0 & 0^* & 5 \\ 4 & 5 & 0^* \end{matrix}$$

FIGURE 2-2
Solution of a 3 × 3 assignment problem.

ond than in the first. Since this must be true of every assignment, it follows that the minimum-cost assignment in the second matrix must have the same pattern (that is, the stars will be in the same places) as the minimum-cost assignment in the first matrix. Hence, if we can solve the second problem we can solve the first one; that is, the cheapest possible assignment in the second problem must be the same as the cheapest possible assignment in the first problem, the only difference being that its total cost is less by one.

Proceeding in the same fashion, we go from the second matrix to the third by subtracting 2 from each element in the first row and by subtracting 3 from each element in the second row, denoted by −5 on the arrow. Once again, if we can solve this third problem, we can solve the second and therefore the first, but the total cost will be less by 6. Now, by the same logic by which we can subtract from rows, we can subtract from columns. We therefore proceed to copy the first column (that is, we subtract 0 from each element of this column), in the fourth matrix. Subtract 1 from the second column (to be rigorous we should say we subtract 1 from each element of the second column, but we use the shorter wording), and copy the third column. Once more, if we can solve this problem, the same solution will apply to the first problem, the only difference being a difference in cost of 7.

We now call your attention to the starred assignment in this right-hand matrix. Note that its total cost is zero. Since there are no negative numbers in this matrix, no assignment can possibly have a cost lower than zero, and therefore this assignment must represent the solution of this problem. It follows that the corresponding assignment in the original problem must be the solution to that problem, and have a cost 7 more than 0 (namely, 7); and if you will add up the cost of the three corresponding elements in the first matrix, you will see that this assignment does indeed have a cost of $2 + 4 + 1 = 7$.

It is now clear why we subtracted the amounts that we did. We subtracted just enough to generate at least one zero in each row (or column), but not enough to generate any negative numbers. It should also be clear that we have demonstrated what the mathematician calls a theorem, namely: The pattern of optimal assignment of an assignment problem is not altered by subtracting a constant from every element of any row or from every element of any column. We have also developed a first approach to an algorithm for solving assignment problems: Subtract from every row the smallest element in that row,

$$\begin{matrix}2&6&9\\3&7&8\\5&4&1\end{matrix} \xrightarrow{-6} \begin{matrix}0&4&7\\0&4&5\\4&3&0\end{matrix} \xrightarrow{-3} \begin{matrix}0&1&7\\0&1&5\\4&0&0\end{matrix} \xrightarrow{+1} \begin{matrix}0&1&7\\0&1&5\\5&1&1\end{matrix} \xrightarrow{-2} \begin{matrix}0^{*}&\checkmark 0&6\\ \checkmark 0&0^{*}&4\\5&0&\checkmark 0^{*}\end{matrix}$$

FIGURE 2-3
A more difficult 3×3 assignment problem.

and then subtract from every column the smallest element in that column. Then select an assignment whose total cost is 0. Unfortunately, life is cruel, and this algorithm does not always work, as we shall now demonstrate.

The assignment problem on the left of Figure 2-3 (by now you will recognize that any square matrix can represent an assignment problem) is identical to our first problem except that we have permuted (that is, interchanged) the bottom two numbers in the second column. We proceed as before, subtracting 2 from the first row, 3 from the second row, and 1 from the third row, giving us the second matrix. Subtracting 3 from the second column gives us the third matrix. Now we have an unforeseen difficulty. There are lots of zeros in this matrix (four to be exact), including at least one in each row and in each column. Thus, we cannot subtract any more without generating negative numbers, and this would destroy the basis of our algorithm; remember that the zero-cost assignment which we found was demonstrably the lowest cost assignment only because there were no negative numbers in the matrix. Nonetheless, in this matrix there is no zero-cost assignment. By "zero cost" we mean "zero relative cost"; the numbers in our derived matrices are all relative-cost numbers.

Let us now be specific as to precisely what constitutes an assignment. It consists of selecting precisely n elements (in this case $n = 3$) such that exactly one of these elements is in each row and exactly one of them is in each column. In the third matrix there exists no such assignment consisting entirely of zeros. At this point it is convenient to introduce a new word: "independence." We shall say that a set of zeros is *independent* if no two zeros in the set are in the same row, and also if no two zeros in the set are in the same column. Note that independence is a property of the set and is therefore a mutual property; "an independent zero" is meaningless unless we give it a special meaning. And we shall do so: When we speak of "an independent zero" we shall mean one of the zeros in a set of independent zeros.

Using this new term, we can say that the difficulty with the third matrix is that it contains no set of three independent zeros, although there are several different sets of two zeros which are independent. We shall resolve this difficulty by a clever trick. Note that the same logic which permits us to subtract from rows or columns also permits us to add to rows or columns; in fact, as noted in Section 0.1.2, subtraction is a special case of addition, and vice versa. Thus, subtraction of a negative number and addition of a positive number are identical. We therefore proceed to add 1 to (subtract −1 from) the third row, yielding the fourth matrix.

What good did this do us? By adding to the third row we have made it possible to subtract from both the second column and the third column, and by doing so we will subtract more than we have added and generate new zeros. This operation yields the fifth matrix, which, it should be noted, has six zeros. More importantly, it has the set of zeros marked by stars, which are independent. Furthermore, we note a new phenomenon which we shall see again many times, namely, the existence of alternate-optimal solutions or *alternate optima*. That is, the set of independent zeros marked by checks represents a different solution to the problem, which has the same zero cost. The net total subtracted in going from the first to the last matrix is 10 (we have subtracted 11 and added 1), and therefore either of these solutions, when transferred to the original matrix, must show a cost of 10.

The algorithm which we have developed for solving assignment problems thus proceeds as follows:

Step 1 Subtract from every row its smallest element; subtract from every column its smallest element. Go to step 2.

Step 2 If there exists a set of n independent zeros, go to step 3; if not, go to step 4.

Step 3 Star the set of independent zeros, transfer the stars to the original matrix, compute the cost of the assignment as the net total subtracted or as the sum of the starred elements in the first matrix (or better, by both methods, as a check). Stop.

Step 4 Judiciously add to certain rows and columns and subtract from other rows and columns in such a way as to generate new independent zeros. Go to step 2.

Obviously this algorithm would not be very good in practice. The first word in step 4 indicates trouble, but of course step 2 is also ambiguous. In a large matrix, how is one to find out whether or not there are n independent zeros? This is a difficult problem, as we shall see from the following example.

2.1 SOLVING A LARGE EXAMPLE

The unfriendly looking object in the upper left of Figure 2-4 you will immediately recognize as a 10×10 assignment problem. We give you a hint: There are no alternate-optimal solutions to this problem; thus there are more than $3\frac{1}{2}$ million possible different assignments, and one of these is cheaper than all the others. We think you will recognize it as a nontrivial problem to determine this optimum assignment.

We know how to start. We subtract 1 from rows 2, 4, and 8 and 2 from row 9 to yield the matrix in the upper right, and then we subtract 1 from column

1	0	6	8	2	3	6	5	4	9
1	5	9	1	3	7	6	1	4	2
9	0	3	4	4	9	4	3	1	4
4	3	6	2	4	3	6	1	3	9
9	0	9	1	9	7	6	4	7	0
5	3	6	8	7	8	6	5	0	9
2	5	9	3	0	5	9	9	7	9
1	9	9	9	1	5	2	4	7	5
9	4	3	2	3	8	8	2	6	4
6	6	2	9	6	4	0	2	0	7

$\xrightarrow{-5}$

1	0	6	8	2	3	6	5	4	9
0	4	8	0	2	6	5	0	3	1
9	0	3	4	4	9	4	3	1	4
3	2	5	1	3	2	5	0	2	8
9	0	9	1	9	7	6	4	7	0
5	3	6	8	7	8	6	5	0	9
2	5	9	3	0	5	9	9	7	9
0	8	8	8	0	4	1	3	6	4
7	2	1	0	1	6	6	0	4	2
6	6	2	9	6	4	0	2	0	7

$\downarrow -3$

1	0*	5	8	2	1	6	5	4	9
0*	4	7	0	2	4	5	0	3	1
9	0	2	4	4	7	4	3	1	4
3	2	4	1	3	0	5	0	2	8
9	0	8	1	9	5	6	4	7	0
5	3	5	8	7	6	6	5	0	9
2	5	8	3	0	3	9	9	7	9
0	8	7	8	0	2	1	3	6	4
7	2	0*	0	1	4	6	0	4	2
6	6	1	9	6	2	0	2	0	7

$\longleftarrow$

1	0*	5	8	2	1	6	5	4	9
0*	4	7	0′	2	4	5	0	3	1
9	0	2	4	4	7	4	3	1	4
3	2	4	1	3	0*	5	0	2	8
9	0	8	1	9	5	6	4	7	0*
5	3	5	8	7	6	6	5	0*	9
2	5	8	3	0*	3	9	9	7	9
0	8	7	8	0	2	1	3	6	4
7	2	0*	0	1	4	6	0	4	2
6	6	1	9	6	2	0*	2	0	7
|	|	|		|	|	|		|	|

FIGURE 2-4
A larger assignment problem (upper left) and the start of the solution process.

3 and 2 from column 6 to yield the matrix in the lower right. This matrix has 18 zeros. We first ask whether any 10 of them are independent. This is not an easy question to answer, since there are more than 40,000 different ways in which we can choose 10 zeros out of 18. And if the answer to this question should be no, it might be even harder to decide where we should add and subtract in order to generate new independent zeros.

The solution to this problem was developed by an American mathematician named Harold Kuhn, based on a theorem first stated by the German mathematician Dines Koenig, and first proved by the Hungarian mathematician E. Egerváry.

Koenig's Theorem is as follows: In any matrix consisting of elements of two types, the maximum number of independent elements of any one type is equal to the minimum number of lines drawn through rows and columns which cover all elements of that type.

If it is not immediately evident to you that this theorem solves our problem, do not despair. Koenig's Theorem applies to any matrix whose elements can be classified into two types. The two classes of elements in which we are interested are zeros and nonzeros; that is, for the moment we are seeking a set of independent zeros, and an element greater than 0 is not of interest to us, and we don't care whether it's 1 or 27. We are interested in the maximum number of independent zeros in the matrix. Koenig's Theorem essentially tells us to ignore this problem for the moment and solve a different problem which has the same answer. Specifically, we are to seek the smallest number of vertical and horizontal lines (diagonal lines are not permitted) which cover all the zeros. This number is equal to the number that we seek. Thus, if we look at the right-hand matrix of Figure 2-2, we note that it has three independent zeros, and we also notice that there is no way of covering all four of the zeros of this matrix with less than three lines (remember diagonal lines are not permitted). There are numerous ways in which they can be covered with three lines; for example, we could use a line through each column. On the other hand, if we look at the third matrix of Figure 2-3, we see that it has a maximum of two independent zeros, and we also note that by drawing a line through the first column and a line through the third row we can cover all four of the zeros. We now formally repeat *Koenig's Theorem* as specialized for our interest: The maximum number of independent zeros in any matrix is equal to the minimum number of lines (through rows or columns) which cover all zeros in the matrix.

After we've found the set of covering lines, what good does it do us? By adding to each covered row and each covered column, we will eliminate all zeros from the matrix, because all the zeros were covered. Note that we will have added less than n, because there are less than n independent zeros and therefore less than n covering lines. Finally, then, we can subtract from all n columns without creating negative numbers, thus subtracting more than we've added. This is essentially what we have done in Figure 2-3. The third matrix of Figure 2-3 is reproduced at the left of Figure 2-5, where it is shown that all the zeros may be covered by two lines. Adding 1 to the covered row and 1 to the covered column, we arrive at the second matrix of Figure 2-5, and then subtracting from every column we arrive at the last matrix of Figure 2-5. The last matrices of Figures 2-3 and 2-5 are identical, the only difference in the way we have achieved these matrices being that, in Figure 2-5, we added to the first column and then subtracted from it again. Thus, the following two sets of instructions are identical: "Add to all the covered rows and columns and then subtract from every column"; and "add to the covered rows and subtract from the uncovered columns." The first set of instructions explains how we utilize

$$\begin{array}{ccc} 0 & 1 & 7 \\ 0 & 1 & 5 \\ 4 & 0 & 0 \end{array} \xrightarrow{+2} \begin{array}{ccc} 1 & 1 & 7 \\ 1 & 1 & 5 \\ 6 & 1 & 1 \end{array} \xrightarrow{-3} \begin{array}{ccc} 0 & 0 & 6 \\ 0 & 0 & 4 \\ 5 & 0 & 0 \end{array}$$

FIGURE 2-5
An alternative presentation of the last three matrices in Figure 2-3.

Koenig's Theorem; the second set is, however, easier to use in practice.

Thus, our algorithm now reads:

Step 1 Subtract from every row its smallest element; subtract from every column its smallest element. Go to step 2.

Step 2 Find the minimal set of lines which cover all zeros; if there are n lines, go to step 4; if there are less than n lines, go to step 3.

Step 3 Add to the covered rows and subtract from the uncovered columns. Remove the lines. Go to step 2.

Step 4 Star the n independent zeros; transfer the stars to the original matrix; compute the cost; stop.

There are a couple of things wrong with this. In the first place, we haven't really told you how to carry out either step 2 or step 4. In the second place, we have asserted Koenig's Theorem, but we haven't proved that it is true; and you should at this point quite properly doubt it (as we have already adjured you, you should never believe anything we say unless we prove it). The two problems are related. When Kuhn studied this problem about 1950, he found that Egerváry had proved Koenig's Theorem, but his proof was what mathematicians call an "existence theorem"; that is, he had proved that such a set of covering lines must exist, but he didn't show how to find them. We shall now prove Koenig's Theorem constructively; that is, we shall prove that the set of covering lines exists by constructing it. This constructive proof will essentially constitute an algorithm for the assignment problem.

In going through this proof we are wearing two hats. Wearing one hat, we are mathematicians worrying about independent zeros (which will be starred) and lines which cover zeros. Wearing the other hat we will be a decision maker assigning workers to jobs and then trying to find appropriate reassignments. We will eventually wish to have a computer do this clerical work for us. A computer is fast and accurate, but the one thing that it is not is clever; therefore, we must avoid being clever—which means that we must have an unambiguous rule (which may be arbitrary) for choosing the row and column in which to work. We will use here, and throughout this book, the following arbitrary rule, called the *northwest-corner rule*, for resolving all such choices: Whenever there is a choice between columns, choose the leftmost available column; whenever there is a choice between rows, choose the topmost available row.

Let us then proceed to prove Koenig's Theorem, and incidentally to solve this assignment problem and assign the workers to the jobs in the best possible way. Using the matrix in the lower right of Figure 2-4 we start by trying to find as many independent zeros as possible, and with this in mind we start with the first available column (which, according to our rule, is the leftmost column) and choose the topmost available row (that is, the row nearest to the top which contains a zero), which happens to be the second row. We star this zero. Thus, the zero in (2,1) is our first candidate for a set of independent zeros. Similarly, we star (1,2), that is, the zero in the first row and second column, and (9,3). When we come to column 4, however, the 0 in the second row, that is, (2,4), is not available, because this 0 and the starred 0 in the first column are not independent. If you are clever, it may now occur to you that we could have done better by picking the 0 in (8,1) instead of the 0 in (2,1), but we have promised not to be clever (and it turns out that the algorithm will find this for us).

Since the second row in the fourth column is not available, we continue downward and try the ninth row in the fourth column; this also is not available (why?). Thus, we were unable to find an independent 0 in the fourth column, and we continue on to the fifth. We remind you again that we are being imprecise here in our use of the phrase "an independent zero," since independence is a mutual property. We should have said, "We are unable to find a 0 in the fourth column which can be added to the set of three independent zeros already identified in such a way as to constitute a set of four independent zeros." But we believe that you will prefer with us the more concise statement.

Continuing onward in this fashion we complete the set of stars as indicated in the lower left matrix and note that we have found eight independent zeros (did you check to make sure that we had not cheated?). The question now is: Do there exist nine independent zeros in this matrix? Koenig's Theorem tells us that one of two things must be true: Either the maximum number of independent zeros in this matrix is eight, in which case we can cover all the zeros in the matrix with eight lines, or the maximum number of independent zeros in this matrix is at least nine, in which case we can find another independent zero. We shall now proceed in such a way as to attempt to find either one of these two things, that is, to find either a set of eight lines that cover all the zeros or a ninth independent zero.

We begin by drawing eight lines which cover many of the zeros, and an obvious way to do this is to cover every column which contains a starred zero. This has been done in the lower left matrix, but instead of drawing the lines all the way down through the matrix, which would make it difficult for you to read the numbers, we have drawn a vertical line at the bottom of the column to indicate covering; this will also make it easier for you when you do the problem, since you will sometimes have to erase such lines. We will now attempt to cover additional zeros with the same number of lines. In order to do this we will cover a row and simultaneously uncover a column, and each time we perform such a pair of operations, we will cover at least one new zero while making sure that all the starred zeros remain covered.

The algorithm proceeds as follows. Look for an uncovered zero. Working in the leftmost available column (column 4, since the first three columns are covered and there cannot be an uncovered zero in a covered column), we find a zero in (2,4) and put on it a different kind of identifying mark, a little diagonal line, which mathematicians call a *prime*. This also is indicated in the lower left matrix of Figure 2-4.

To make sure that you are with us, we give you a clean matrix in the upper left corner of Figure 2-6 and ask you to put in the stars, primes, and coverings with pencil. Do it lightly, because you will have to erase some of them. After you have placed eight stars, eight covering lines, and one prime, your picture should look just like ours.

Now remember what we are trying to do with our algorithm. We are either going to find a ninth independent zero, or we are going to cover all 18 zeros with eight lines. Toward this end, we ask you to cover the row of the primed zero (row 2) with a horizontal line on the right of the matrix and uncover the column of the starred zero in this row (column 1) by erasing the line below this column. Note what we have accomplished. We still have only eight covering lines, but now in addition to having all the starred zeros covered, we also have a primed zero, and this is also covered. As we proceed we will continue to uncover columns, but we will never uncover a starred or primed zero (we may uncover its column, but then its row will be covered). We should also interpret what we have accomplished in terms of the decision problem: The starred zeros represent tentative assignments; the primed zero represents a candidate for reassignment. The fact that a column was uncovered meant that this job was considered for different assignment.

We now return to the algorithm. Look for an uncovered zero. The leftmost available column is now the first column (which we have just uncovered); the topmost zero (in row 2) is not available, because its row is covered. The next zero (in row 8) is available, and we prime it. We now note that we cannot proceed as before, because there is no starred zero in the eighth row. However, this is good: It means that we have found our new independent zero, because the primed zero represents a candidate for reassignment, and the lack of a star in this row means that this candidate may be reassigned. This will then lead to a *chain* of reassignments.

The algorithm proceeds as follows: "There now exists a unique chain (which may contain just one zero, or three, or five, . . .) as follows: Go vertically from the uncovered primed zero to the starred zero in this column (if any); from there go horizontally to a primed zero; from there go vertically to a starred zero in this column (if any); and so forth. The chain ends when you arrive at a primed zero (which might be the first one), which has no starred zero in its column. Go through this chain so you know where it is, then go back to the beginning of the chain (that is, the uncovered primed zero), and go through the entire

1	0	5	8	2	1	6	5	4	9
0	4	7	0	2	4	5	0	3	1
9	0	2	4	4	7	4	3	1	4
3	2	4	1	3	0	5	0	2	8
9	0	8	9	7	5	6	4	7	0
5	3	5	8	7	6	6	5	0	9
2	5	8	3	0	3	9	9	7	9
0	8	7	8	0	2	1	3	6	4
7	2	0	0	1	4	6	0	4	2
6	6	1	9	6	2	0	2	0	7

→

1	0*	5	8	2	1	6	5	4	9
0	4	7	0*	2	4	5	0	3	1
9	0	2	4	4	7	4	3	1	3
3	2	4	1	3	0*	5	0	2	8
9	0	8	1	9	5	6	4	7	0*
5	3	5	8	7	6	6	5	0*	9
2	5	8	3	0*	3	9	9	7	9
0*	8	7	8	0	2	1	3	6	4
7	2	0*	0	1	4	6	0	4	2
6	6	1	9	6	2	0*	2	0	7
|	|	|	|	|	|	|		|	|

↓

1	0*	5	8	2	1	6	5	4	9	
1	5	8	1*	3	5	6	1′	4	2	—
9	0	2	4	4	7	4	3	1	4	
4	3	5	2	4	1*	6	1′	3	9	—
9	0	8	1	9	5	6	4	7	0*	
5	3	5	8	7	6	6	5	0*	9	
2	5	8	3	0*	3	9	9	7	9	
0*	8	7	8	0	2	1	3	6	4	
8	3	1*	1′	2	5	7	1	5	3	—
6	6	1	9	6	2	0*	2	0	7	
|	|			|		|		|	|	
					↓ −4					

+3 ←

1	0*	5	8	2	1	6	5	4	9	
0	4	7	0*	2	4	5	0′	3	1	—
9	0	2	4	4	7	4	3	1	4	
3	2	4	1	3	0*	5	0′	2	8	—
9	0	8	1	9	5	6	4	7	0*	
5	3	5	8	7	6	6	5	0*	9	
2	5	8	3	0*	3	9	9	7	9	
0*	8	7	8	0	2	1	3	6	4	
7	2	0*	0′	1	4	6	0	4	2	—
6	6	1	9	6	2	0*	2	0	7	
|	|			|		|		|	|	

FIGURE 2-6
Continuation of the solution of the large assignment problem.

chain, changing each prime to a star and erasing all stars which were already there." This means that we should change the prime in (8,1) to a star, erase the star in (2,1), and change the prime in (2,4) to a star. Now erase all row coverings, and cover every column which contains a starred zero.

If you have done all this correctly, your diagram in the upper left of Figure 2-6 should look exactly like ours in the upper right. But you will have to continue to work in the upper left because you can't erase in our printed diagram.

We note that we now have nine independent zeros, one more than before; the question is: Does there exist a set of 10 independent zeros in this matrix? Koenig's Theorem tells us that one of two things must be true: either the maximum number of independent zeros in this matrix is nine, in which case we can cover every zero in the matrix with nine lines, or the maximum number of independent zeros in the matrix is 10, in which case we can find a new independent zero.

We proceed as before. The leftmost available column (in fact the only available column) is the eighth, and we prime the topmost available zero in (2,8). Find the starred zero in the same row, which is in the fourth column, cover the row (the second row), and uncover the column of the starred zero (the fourth column). The leftmost available column is now the fourth, the topmost uncovered zero in this column is in the ninth row; we prime (9,4), cover row nine, and uncover column three. The next uncovered zero you will find is in (4,8); prime it, cover row four and uncover column six (the column that contains the starred zero in the fourth row). Your diagram should now be exactly like ours in the lower right of Figure 2-6. At this point when you look for an uncovered zero, a new phenomenon appears: You will not be able to find one, because we have achieved something remarkable; we have covered all 18 of the zeros with just nine lines, six vertical and three horizontal. This, you remember, was our objective; we have proved Koenig's Theorem constructively; that is, we have constructed a set of nine independent zeros, which are identified in our matrix in Figure 2-6 by stars, and we have constructed a set of nine lines which covers each of the 18 zeros in the matrix. The first part proves that there are at least nine independent zeros in the matrix, because we have found them; and the last part proves there are at most nine independent zeros in the matrix, because every zero in the matrix lies in one of the nine lines, and you will remember that if zeros are to be independent, they must lie along different lines. Thus, we have proved Koenig's Theorem (at least for this matrix).

We have also found how to generate new zeros (which, if we are lucky, will include a new independent zero). Namely, we will add to the three covered rows and the six covered columns, thus adding a total of nine, and then subtract from all columns, thus subtracting a total of 10. However, as we said before, this involves duplicating some work; for example, there is no point in adding to the first column which is covered and then subtracting from it again. You remember our short cut. We will add only to the three covered rows and then subtract from the four uncovered columns, which has the same effect. This is done in our next two matrices: the lower left in Figure 2-6 and the upper left in Figure 2-7. Note in the first of these (after we have added to the three covered rows) that there are no zeros left in the uncovered columns. This is necessary if we are to subtract from these columns without generating negative numbers. Note also that in this intermediate stage, some primes and stars are attached to

1	0	4	7	2	0	6	4	4	9
1	5	7	0	3	4	6	0	4	2
9	0	1	3	4	6	4	2	1	4
4	3	4	1	4	0	6	0	3	9
9	0	7	0	9	4	6	3	7	0
5	3	4	7	7	5	6	4	0	9
2	5	7	2	0	2	9	8	7	9
0	8	6	7	0	1	1	2	6	4
8	3	0	0	2	4	7	0	5	3
6	6	0	8	6	1	0	1	0	7

1	0*	4	7	2	0′	6	4	4	9
1	5	7	0*	3	4	6	0′	4	2
9	0′	1	3	4	6	4	2	1	4
4	3	4	1	4	0*	6	0′	3	9
9	0	7	0′	9	4	6	3	7	0*
5	3	4	7	7	5	6	4	0*	9
2	5	7	2	0*	2	9	8	7	9
0*	8	6	7	0	1	1	2	6	4
8	3	0*	0′	2	4	7	0	5	3
6	6	0′	8	6	1	0*	1	0	7

1	0	6	8	2	3*	6	5	4	9
1	5	9	1*	3	7	6	1	4	2
9	0*	3	4	4	9	4	3	1	4
4	3	6	2	4	3	6	1*	3	9
9	0	9	1	9	7	6	4	7	0*
5	3	6	8	7	8	6	5	0*	9
2	5	9	3	0*	5	9	9	7	9
1*	9	9	9	1	5	2	4	7	5
9	4	3*	2	3	8	8	2	6	4
6	6	2	9	6	4	0*	2	0	7

+9 / −9

1	0	4	7	2	0*	6	4	4	9
1	5	7	0*	3	4	6	0	4	2
9	0*	1	3	4	6	4	2	1	4
4	3	4	1	4	0	6	0*	3	9
9	0	7	0	9	4	6	3	7	0*
5	3	4	7	7	5	6	4	0*	9
2	5	7	2	0*	2	9	8	7	9
0*	8	6	7	0	1	1	2	6	4
8	3	0*	0	2	4	7	0	5	3
6	6	0	8	6	1	0*	1	0	7

FIGURE 2-7
Completion of the solution of the large assignment problem.

numbers other than zeros, but these are all in uncovered columns, and by the time we get to Figure 2-7 all the primed and starred zeros have been returned. It is possible to lose a zero [for example, in the adding and subtracting, the zero in (2,1) disappeared]. On the other hand, we are bound to get some new zeros, such as the one in (10,3), which is the first one we shall deal with in the next stage of the algorithm. We have left the matrix in the upper left of Figure 2-7 clean for you to work on, and we ask you to copy the nine stars, three primes, and nine covering lines lightly in pencil onto this diagram. If you've done it properly, the stars will be on independent zeros, and there will be just one prime in each covered row (why?).

Working then in the leftmost available column (column three) we note that (9,3) is unavailable because its row is covered, but (10,3) is available. We prime it, locate the starred zero in (10,7), cover the tenth row, and uncover the seventh column.

By this time you should be able to figure out what to do next. See if you can do it before you read on. Which is the leftmost available column now? Which zero do you pick? What mark do you put on it? What row do you cover? What column do you uncover? Write your answers on a piece of scratch paper before you read ours. The answers are the fourth column; (5,4); prime; the fifth row; the tenth column. And the next one is: Prime the zero in (9,6), cover the first row, uncover the second column.

You may well ask: How long can this go on? It's a good question. We could have solved this problem in the first place by trying $3\frac{1}{2}$ million different assignments, and if this method is going to take $3\frac{1}{2}$ million steps, we won't have saved anything. It is a requirement of a good algorithm not only that it *converge* (that it must ultimately come to an answer), and converge to the right answer, but also that it do so fairly rapidly. The point is that this algorithm does converge rapidly. Note that we always cover rows and uncover columns; we never go the other way. At this point there are only three columns left to uncover; in three more steps, at the most, we must either find a new independent zero or succeed in covering all the zeros with our nine lines.

Again, before looking, decide for yourself what to do next. Our answer: You work in the leftmost uncovered column, which is the second one, prime the zero in (3,2), and note that there is no starred zero in this row; therefore we are finished—we have found our tenth independent zero. Now remember the "chain" rule. Starting with this most recently primed zero, in (3,2), we go vertically to a starred zero, then horizontally to a primed zero, and continue in this way until the chain comes to an end because we cannot find a starred zero in the same column. Thus, we go from (3,2) to (1,2) to (1,6) to (4,6) to (4,8), and there the chain ends because there is no starred zero in column eight. We show you in the upper right of Figure 2-7 how your diagram should look, and we also mark the chain (but you don't have to do that).

Having established this chain, go back to the beginning (do not attempt to go through the chain backwards); that is, start at (3,2), changing primes to stars and erasing the stars that were already there. Now erase all primes and all coverings; the result is shown in the lower right of Figure 2-7. Check to make sure that the problem is really solved; that is, note that there are exactly 10 stars and that there is exactly one in each row and exactly one in each column. Now transfer the stars back to the original matrix, as in the lower left of the figure, and compute the cost both ways: Note that we subtracted first 5 and then 3 and that we then added 3 and subtracted 4. Thus, the net total subtracted in reaching the zero-cost solution was 9, and the cost in the original matrix must be nine. Adding up the cost of the assigned elements in the lower left matrix, we find that it is indeed nine. If you would like to check the remaining 3,628,799 assignments, you will find that each of them costs 10 or more. We will prove a little later that our solution is the unique-optimal solution.

We have accomplished the following things: We have solved the original assignment problem, we have sort of proved Koenig's Theorem, and we have developed an algorithm for solving all assignment problems. This algorithm was developed by the American mathematician James Munkres (as a modification of Kuhn's "Hungarian method") in the late 1950s. It is remarkable that you now know more about the assignment problem than was known by the best mathematicians in the world just a generation ago. We next state this algorithm formally.

We said that we had "sort of proved" Koenig's Theorem. Actually we have proved it for the particular matrices of Figures 2-4, 2-6, and 2-7 and made it appear reasonable that it is true for all other matrices. But mathematicians have learned through painful experience that to "appear reasonable" is not enough, because exceptions pop up (we shall see some examples of this). Hence, mathematicians demand a high degree of rigor in their proofs. To prove Koenig's Theorem we would have to show that our algorithm always works—that is, that no possible matrix would provide an exception. Munkres has given such a proof; we ask you to carry out part of this proof in Problem 2-7.

2.2 ALGORITHM FOR THE ASSIGNMENT PROBLEM

Step 1 Subtract from every row its smallest element; subtract from every column its smallest element. Keep track of the amounts subtracted. Star the topmost zero in the first column. Go to step 2.

Step 2 In the next column, star the topmost zero, if any, which is in a row having no other star. If this is the last column go to step 3; if not, go to step 2.

Step 3 Cover every column which has a star. If all columns are covered go to step 9; if not, go to step 4.

Step 4 Look (starting in the leftmost uncovered column) for an uncovered zero. If one is found, go to step 5; if none is found, go to step 8.

Step 5 Prime this zero and look for a starred zero in the same row. If one is found, go to step 6; if none is found, go to step 7.

Step 6 Cover this row, uncover the column of the starred zero, and go to step 4.

Step 7 There now exists a unique chain (which may contain just one zero, or three, or five, . . .) as follows: Go vertically from the uncovered primed zero to the starred zero in this column (if any); from there go horizontally to a primed zero; from there go vertically to a starred zero in this column (if any); and so forth.

The chain ends when you arrive at a primed zero (which might be the first one), which has no starred zero in its column. Go through this chain so you know where it is, then go back to the beginning of the chain (that is, the uncovered primed zero), and go through the entire chain, changing each prime to a star and erasing all stars which were already there. Now erase all primes and all row coverings. Go to step 3.

Step 8 Find the smallest uncovered element in the matrix; add this number to the covered rows and subtract it from the uncovered columns (or add it to the covered columns and subtract it from the uncovered rows, or add it to the twice-covered elements and subtract it from the uncovered elements). Keep track of the amounts added and subtracted. Do not change any primes, stars, or coverings. Go to step 4.

Step 9 Erase all coverings. Compute the total cost by adding up the total net amount subtracted. Transfer the stars back to the original matrix and recompute the cost by adding the cost of the assigned elements. If you are not interested in alternate-optimal solutions, stop; if you are, go to step 10.

Step 10 Left as an exercise. We shall talk about alternate-optimal solutions later.

There are two new points here, both in step 8. Remember that our theoretical basis for adding and subtracting was that we were going to add to all covered rows and columns and then subtract from all columns. However, as a shortcut we chose only to add to the covered rows and subtract from the uncovered columns. Clearly there is a duality principle here: We could equally well have added to the covered columns and subtracted from the uncovered rows; we chose (at the bottom of Figure 2-6) to add to the covered rows and subtract from the uncovered columns because there were only three and four of them, respectively, whereas if we'd added to the covered columns and subtracted from the uncovered rows, there would have been six and seven, respectively. One has the alternative of going either way, and you should note that the effect of doing either of these operations is to add to twice-covered elements (that is, those elements which are in a covered row and a covered column), to make no change in those elements that are covered once (that is, either in a covered row and uncovered column, or in a covered column and uncovered row), and to subtract from uncovered elements (that is, those elements which are in both uncovered rows and uncovered columns). This last choice is probably the one that you will find easiest to do when you are actually working the problem out by yourself, since it requires writing the entire matrix over only once instead of twice.

The other point is how much to add and subtract. We have, without explanation, added and subtracted 1, and it has worked in all our examples, because there were always 1s which were not covered and which turned into zeros. However, this might not have happened; we could be working with a problem in which the costs were numbers like \$17.22, in which case it probably wouldn't happen (we chose small integers only to simplify the example). However, if we pick the smallest uncovered element in the matrix and add and subtract this amount, it is clear that we will accomplish our objectives; namely, we will generate at least one new uncovered zero (why?), and we will not generate any negative numbers (why?).

Some hints for doing this algorithm follow. Of greatest importance: Make haste slowly. If you start with Problem 2-1 at the end of this chapter, which is a 7×7 problem, and if you go very slowly and deliberately and carefully, it may take you 30 to 60 minutes (after you've done several of them, of course, you will be able to do a problem of this size in much less time). If you try to do it a little more rapidly, it will take you about 2 hours; and if you try to go through it very quickly, you will probably never finish. We speak to you out of deep and painful experience; nobody can do this kind of thing rapidly the first few times. We strongly advise that you write the numbers in ink and do not attempt to erase them. The primes, stars, and coverings may be put in by pencil, because these may be erased. When you get to step 8, write out a new matrix, in ink. If at any time you find you have made a mistake (for example, if you find two stars in the same row or column, or if there is no prime in the row when you are trying to go through the chain in step 7, or if you simply get thoroughly confused), simply erase all pencil marks and go back to step 2 with this latest matrix. It is not necessary to go back to the beginning of the problem.

2.3 THEORETICAL INTERPRETATION

Consider the following problem:

Given: $$C = \|c_{ij}\|$$

Find: $$X = \|x_{ij}\|$$

Subject to: $$\sum_i x_{ij} = 1 \qquad j = 1, \ldots, n$$

and $$\sum_j x_{ij} = 1 \qquad i = 1, \ldots, n \tag{2-1}$$

and $$x_{ij} = x_{ij}^2 \quad \Big| \quad i, j = 1, \ldots, n$$

To minimize: $$z = \sum_i \sum_j c_{ij} x_{ij}$$

Is it obvious that this is the problem we have just solved? No? It will be.

Look at the first line. It says that we are given a matrix of costs, and this is, of course, just the matrix we started with in the upper left of Figure 2-4 (or

0	0	0	0	0	1	0	0	0	0
0	0	0	1	0	0	0	0	0	0
0	1	0	0	0	0	0	0	0	0
0	0	0	0	0	0	0	1	0	0
0	0	0	0	0	0	0	0	0	1
0	0	0	0	0	0	0	0	1	0
0	0	0	0	1	0	0	0	0	0
1	0	0	0	0	0	0	0	0	0
0	0	1	0	0	0	0	0	0	0
0	0	0	0	0	0	1	0	0	0

FIGURE 2-8
A matrix of x's (a permutation matrix).

Figures 2-2 or 2-3 if you prefer). We are then asked to find a matrix of the same size, of x's. For example, in Figure 2-4 we would have been asked to find a matrix of 100 x's. These x's are subject to certain constraints, indicated in the next three lines. Look first at the last constraint; it says $x_{ij} = x_{ij}^2$ for every possible i from 1 to n and also for every possible j from 1 to n (that is, for all the 100 x's). $x_{ij} = x_{ij}^2$ is a quadratic equation which has two solutions, and as you can see, these solutions are 0 and 1. Thus, we are asked to find 100 x's, each of which is either 0 or 1. Now look at the first constraint (the third line). For each j from 1 to n, that is, for every column, when we add up all the x's in that column (that is, when we take the sum over each value of i) the total must be exactly equal to 1. Now, if we have a column of x's, each of which is either a 0 or 1, and if the sum of the entire column is 1, what do we know? We know that the column contains exactly one 1 and all the rest are 0s. The next line says that for each row the same thing must be true; that is, every row must have exactly one 1, and all the other elements of the row must be 0. When you take all these constraints together, they say that the matrix of x's which we are asked to find must be what we have called a permutation matrix. That is, it must be a matrix consisting entirely of 1s and 0s, with exactly one 1 in each row and exactly one 1 in each column. Now there are $n!$ (remember that means $1 \times 2 \times \cdots \times n$) different $n \times n$ permutation matrices (in this case 3,628,800 of them), and we are asked to find the one which minimizes the last line. This line tells us that after we have our x-matrix, we take each x and multiply it by the corresponding c, then we sum over j (that is, we add up each row to get 10 different sums), and then we sum over i (that is, we sum the sums). This grand total is to be minimized by choosing an appropriate set of x's. For example, Figure 2-8 is an x-matrix that satisfies the constraints (it also happens to be the optimum one). In carrying out the multiplication of c_{ij} (from Figure 2-4 or 2-7) by x_{ij} (from Figure 2-8), indicated in the last line of (2-1), we are asked first to take the sum of the following products for the first row: $1 \times 0 + 0 \times 0 + 6 \times 0 + 8 \times 0 +$

$2 \times 0 + 3 \times 1 + 6 \times 0 + 5 \times 0 + 4 \times 0 + 9 \times 0$. We will do the same thing with each of the other nine rows and then add up these sums.

The x's, then, are simply numbers which are 1 where we have put a star and 0 where we have not put a star. The first constraint says that for every job (for every j) there must be exactly one assignment; the second constraint says that for every worker (every i) there must be exactly one assignment; the last constraint says that a worker must be either assigned or unassigned, you cannot half-assign a worker to a job. Finally, we multiply the costs by 1 where there is an assignment (that is, we count it) and by 0 where there is no assignment (that is, we do not count it), and then we add up these costs. Thus, it should now be obvious that this is the problem we have just solved. If it still isn't obvious, maybe you should start at the beginning of Section 2.3 again, or even go back and read Section 1.5 over.

Now (2-1) turns out to be an extremely difficult problem to solve. The last set of constraints, which are quadratic equations, are nonlinear, and there are 100 of them. In this book we deal only with linear problems, which are comparatively easy; any problem with a large number of nonlinear constraints is almost certain to be extremely difficult. Since we do not know how to solve this problem, we look at a similar problem which we think we probably can solve.

Given: $C = \|c_{ij}\|$

Find: $X = \|x_{ij}\|$

Subject to: $\sum_i x_{ij} = 1 \qquad j = 1, \ldots, n$

and $\sum_j x_{ij} = 1 \qquad i = 1, \ldots, n \qquad$ (2-2)

and $x_{ij} \geq 0 \qquad i,j = 1, \ldots, n$

To minimize: $z = \sum_i \sum_j c_{ij} x_{ij}$

This problem differs from the previous one only in the fifth line. Before, we required that each x be either 0 or 1; now, we require only that it be 0 or positive. When we solve problem (2-2)—remember that the parentheses tell you this is the set of equations (2-2), not a problem at the end of the chapter—the set of x's which minimizes the cost will always be integers. We will prove this in Section 6.4, after we know a little more about linear programming. Meanwhile, we ask you to take it on faith. However, if the x's are integers, the last constraint guarantees that these integers will be not less than 0, and the first two constraints guarantee that they will be not more than 1, so they must be either 0 or 1; therefore the quadratic constraints of (2-1) will be satisifed even if we ignore them.

As we shall see, (2-2) belongs to the great class of problems known as *linear programming problems*; we shall learn a great deal about them in this book, and one of the things we shall find out is that every linear programming problem has a dual problem, and from the solution of one we can always determine the solution of the other. Again, we ask you for the moment to take it on faith that the dual of the problem presented in (2-2) is

Given:	$C = \|c_{ij}\|$	
Find:	$U = \|u_i\|$ and $V = \|v_j\|$	
Subject to:	$u_i + v_j \leq c_{ij} \quad i = 1, \ldots, n; j = 1, \ldots, n$	(2-3)
To maximize:	$w = \sum_i u_i + \sum_j v_j$	

This problem says that we are given the same cost matrix, but now instead of being asked to find 100 x's, we are asked to find 10 u's (a vector of u's, one u for each row), and 10 v's (a vector of v's, one v for each column). There is only a single type of constraint, but there are 100 constraints of this type. Finally, we are asked to maximize the sum of all the u's and v's selected.

Would you believe that this is really the problem we have just solved? The u's represent the amounts we have subtracted from each row; the v's represent the amounts we have subtracted from each column. The constraints say that the total amount that you subtract from an element (that is, u_i, the amount you subtract from its row, plus v_j, the amount you subtract from its column) cannot exeed c_{ij}, the original cost of that element (because otherwise that element would go negative). Subject to these constraints, we wish to subtract as much as possible. Now please note that w, the quantity to be maximized in the problem (2-3), which is the amount subtracted to reach the zero-cost solution, must equal z, the quantity to be minimized in the problem (2-2), which is the cost of the assignment, so that these dual problems have the same numerical solution.

Consider the sophistication of what we have done. We start with a managerial decision problem of assigning workers to jobs in such a way as to minimize the cost. We recognize that this can be modeled as a matrix problem, and so we write out its formal mathematical description as in (2-1). Recognizing that this is a nonlinear (and therefore difficult) problem, we examine the related linear problem (2-2) and are able to show that it has the same solution. Then we recognize that (2-2) is a linear programming problem and has a dual problem which we write as (2-3). Finally, this dual problem is solved through Koenig's Theorem. And what we end up with is the algorithm of Section 2.2, which requires no arithmetic more advanced than simple additions and subtractions, and which can be taught to any clerk, or which can be easily programmed for a digital computer. We would like to stress, however, that this elementary algorithm could not have been developed by inspired search combined with luck without the very sophisticated knowledge of matrix mathematics which was available to people like Koenig, Egerváry, Kuhn, and Munkres. That is, the developers of such algorithms must be sophisticated mathematicians. The users of such algorithms can be people with quite ordinary educational attainments.

2.4 ALTERNATE OPTIMA

We have identified an optimum solution to our 10×10 assignment problem, with a cost of 9. In the process we have proved that no assignment can have a cost less than 9 (how have we proved this?), but we have not yet determined whether or not there are other assignments whose cost is equal to 9. Such alternate optima are of considerable importance in real practice, because two solutions whose cost may be the same according to the model may actually have different values in terms of factors which have not been modeled. Remember that the model is never a complete picture of the real world; it is an abstraction which may be very useful. But, because it is imperfect, it may also be very misleading. For example, c_{37} (that is, the c_{ij} which describes the performance of the third worker on the seventh job) might turn out in real life to be much higher than we thought, because the chosen assignment places the third worker, a young man, across from a very attractive young lady, which prevents him from keeping his mind on his work. Such an interaction, which represents a nonlinearity, was not taken into account when we constructed the model of Figure 2-4.

In our search for alternate optima you may work on the lower right matrix of Figure 2-7 (or the upper left if you prefer). It has 20 zeros, of which 10 are starred. Could we find other sets of 10 independent zeros, as we did on the right of Figure 2-3? Look at the third row (which represents the third worker), and note that there is only one job (namely, the second), which he can do at zero cost. If we are to find an optimal solution to this problem (the problem as portrayed in Figure 2-7), it must have a total cost of 0, and therefore the second job must be assigned to the third worker; that is, we must use the 0 in (3,2). It follows that the second job can't be assigned to any other worker, and more particularly, it can't be assigned to either the first worker or the fifth; that is, we cannot utilize the 0s in (1,2) or (5,2). Since these 0s cannot be used, they might as well not be there, and you scratch them out by drawing an $\times$ through them like this $\not{0}$ (see also Figure 2-9). But note that because we have scratched out the 0 in (1,2) there remains only one 0 in the first row—namely, (1,6)—and so you can scratch out any other 0s in this sixth column—specifically, the one in (4,6). This makes the 0 in (4,8) the only 0 in the fourth row, and so you should scratch out the other 0s in the eighth column.

As we have mentioned before, there is complete duality in assignment problems between rows and columns, so if we find a zero which is the only one in its column, such as (8,1), we may delete the zeros in its row. Thus you can delete (8,5), and because there is only one 0 in the seventh column you can

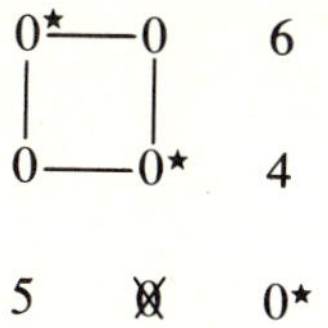

FIGURE 2-9
Alternate-optimal solutions in a 3 × 3 matrix.

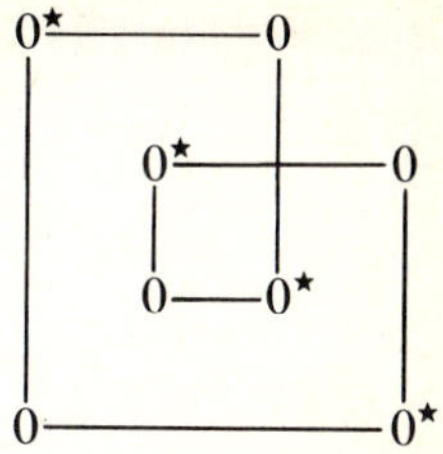

FIGURE 2-10
Alternate-optimal solutions in a larger matrix.

delete (10,3) and (10,9). Finally, you will note that there remains only one zero in the second row, and so you may cross out the other 0s in the fourth column, namely, (5,4) and (9,4).

Examining what you have done, you will see that there are only ten 0s left in the matrix and that all these 0s are starred. What we have proved is that there is no zero (other than these 10) which may be used in any optimal solution and therefore that this particular assignment is the unique-optimal solution.

Of course it need not always work out like this. Even in the simple 3 × 3 case of Figure 2-3 we found that there were two sets of alternate-optimal solutions. If we eliminated zeros from that matrix as above, we would reach the situation of Figure 2-9. Any zeros that remain after the deletion is complete are probably members of alternate-optimal sets. To determine such sets, start at any one of the zeros which is not starred, and draw a chain (as in step 7 of our algorithm), continuing in this case until the chain returns to the starting zero. Such a *closed chain* is shown in Figure 2-9. If all the starred zeros in such a closed chain are deleted, and all the unstarred zeros are added, a new set of independent zeros will be obtained. Note that the chain does not have to include all the zeros in the matrix. A more complicated closed chain is shown in Figure 2-10, which might be a portion of a larger matrix. The starred zeros which are not members of this chain (not shown in Figure 2-10) constitute an optimal solution when taken together either with the starred zeros in the chain or the unstarred zeros in the chain. Finally, in Figure 2-11 we show a situation involving multiple optima; in this case there are six different ways of choosing three independent zeros from the nine shown.

0*	0	0
0	0	0*
0	0*	0

FIGURE 2-11
Multiple alternate-optimal solutions.

2.5 OPTIMIZATION AND SUBOPTIMIZATION

We call your attention to element (9,3) in our final solution (Figure 2-7). It says that the cost of putting the ninth worker on the third job is 3. Now this is not the best worker for the job; the tenth worker can do it less expensively. Nor is this the job that the ninth worker is best at; it's cheaper if you put that worker on jobs 4 or 8. Nonetheless, if you look at this figure you will see that you must put the ninth worker on the third job if you are to achieve the cheapest possible solution to the overall problem.

This exemplifies a very important principle: If you attempt to go row-by-row in solving such a problem as this, that is, if you choose one worker at a time and assign that worker to the best possible job, you will be guilty of the ultimate sin of *suboptimization*. That is, you will have optimized individual subsystems (in this very simple example each worker or each job might be considered a subsystem), and you will not have achieved an overall system optimum. This is really the whole point of a book such as this one. In any kind of complex system there are interactions, and they must be taken into account when optimizing the system. Our primes and stars and coverings did just that—they represented special types of interactions. If you assign one worker to a particular job, then you can't assign another to the same job, and this in turn says something about other jobs that worker might do.

In more complicated systems, of course, the interactions are more complicated, and we shall have to learn more complicated ways of dealing with them. We shall be trying throughout this book to take the *systems viewpoint* in modeling the systems in which we are interested. If the system is too massive or complex, then a systems model may be too difficult to deal with, and we shall be forced to some sort of modification or simplification. But if we attempt to model too small a piece of the system, then we may fall into the opposite fallacy of suboptimization. We shall tread the narrow line between.

2.6 MAXIMIZATION PROBLEM

Suppose now that the numbers in Figure 2-2 or 2-4 really represented performances instead of costs; that is, suppose that the problem (2-1) were the same, except that we said "maximize" in the last line instead of "minimize."

It is hard to tell, when looking at a modification like this, whether it makes the problem much harder or only a little bit harder. In this case it makes it only a little bit harder. We supply the

Algorithm for a Maximization Assignment Problem

Step 0 Multiply every element of the matrix by −1 (that is, change the sign of every element in the matrix). Go to step 1 of the assignment algorithm. Note that when you are asked to locate and subtract the smallest element of a row, it will be the most negative of the negative numbers.

What we have asserted is that, to change a maximization problem into a minimization problem (or vice versa), you simply multiply the objective function by -1. *Objective function* is the name given to whatever quantity we wish to maximize or minimize in an optimization problem. This multiplication by -1 is a very useful trick, and we shall use it repeatedly. It is valid for the following reason: When we say minimize a certain sum, what we really mean is choose that set of x's which gives the smallest sum. Note that there are about $3\frac{1}{2}$ million sets of x's which satisfy the constraints for our problem of Figure 2-4, and each one of these sets of x's gives a particular sum; so conceptually what we are being asked to do is equivalent to writing down all those $3\frac{1}{2}$ million sums and picking the smallest one. If we multiply each of the $3\frac{1}{2}$ million sums by -1, and then chose the largest one, we would clearly be picking the same one and, therefore, clearly picking the same set of x's, which is what we are after.

Another way of stating what we have just asserted is that there is no real distinction between a profit and a cost except that one is the negative of the other; reducing cost by \$1.00 is the same as increasing profit by \$1.00. Thus, minimizing cost is the same as maximizing profit. Thus, throughout this book we shall choose to maximize or minimize, whichever is more convenient for the algorithm we are developing. Then, if the problem happens to go the other way, we shall simply multiply the objective function by -1.

If you solve such a problem (for example, Problem 2-5*b*), the dual variables will be the amounts added to the (initially negative) c_{ij}'s to obtain the zero-cost solution with all c_{ij}'s nonnegative.

2.7 NONSQUARE ASSIGNMENT PROBLEM

We have assumed that the number of workers and the number of jobs is the same; but we might have 10 workers to assign to eight jobs so that the cost matrix is 10×8 and has 80 elements.

The solution of this problem is mathematically easy, although, as we shall see, there can be practical difficulties. What we do is square up the matrix by adding two *dummy* columns for imaginary jobs. If you like, these jobs are not imaginary, but are real jobs of loafing; that is, two of the workers will not be assigned at all—so they will be assigned to the dummy jobs. Now we fill these columns with zeros, and we have a regular assignment problem which we know how to solve. Thus, the mathematics involved in converting this to a soluble problem is easy.

The difficulties in practice are illustrated if we have 10 jobs and only eight workers. In this case, two of the jobs will have to go undone. Obviously, our method of handling this is to add two dummy rows to the problem. These represent workers who work so inefficiently that they don't do the job at all. Now the problem is: What numbers should we put in these rows? It is clear that the two rows must be identical to one another, but for each job we must put in the cost of leaving that job undone, and this may be very difficult to determine. This example demonstrates a fundamental truth about linear programming: It is often far more difficult to find the numbers that go into the matrix in a real problem than it is to solve the problem after we have the matrix.

2.8 THE ASSIGNATION PROBLEM

Consider the following problem: We have n men ($i = 1, \ldots, n$) and n women ($j = 1, \ldots, n$). We define a quantity $a_{ij}^{\rightarrow}$ which represents the value that the ith man places on spending his time with the jth woman; we define a quantity $a_{ij}^{\leftarrow}$, the value that the jth woman places on spending her time with the ith man. Finally we take the arithmetic mean of these two quantities

$$a_{ij} = \frac{a_{ij}^{\rightarrow} + a_{ij}^{\leftarrow}}{2}$$

which represents the average happiness resulting from the liaison of the ith man with the jth woman. We wish to find a normative optimization (whatever that means) in terms of the fraction of time x_{ij} which the ith man spends with the jth woman.

There are some constraints. Start with

$$0 \leq x_{ij} \leq 1 \qquad i, j = 1, \ldots, n$$

because of the definition of x_{ij} as the "fraction" of time. People cannot spend less than none of their time or more than all of it. Furthermore,

$$\sum_j x_{ij} = 1 \qquad i = 1, \ldots, n$$

because, after all, each man has to spend his time with somebody, or at least we ask you to assume that for the duration of this problem. Similarly, and by symmetry,

$$\sum_i x_{ij} = 1 \qquad j = 1, \ldots, n$$

Finally our objective function is to maximize

$$\sum_i \sum_j a_{ij} x_{ij}$$

which represents total happiness.

Now we do not assert this is an easy problem in practice; it is reported that one research group sent out investigators to determine the numerical values of the a_{ij}'s and that they never came back. Nevertheless, we have a theorem that asserts that the optimal values of the x_{ij}'s (when found) will be integers. Thus, the optimal solution to this problem is monogamous.

But this is not the whole story; there are $n!$ different monogamous solutions to this problem, and while it is true that theory tells us that one of these monogamous solutions maximizes total happiness, the same theory tells us that a different monogamous solution minimizes total happiness, and it is by no means clear how one is to determine which solution we are closer to.

This completes our presentation of all that is known about the assignation problem. We thought you would be interested in the theoretical insights that can sometimes be obtained without actually going through a numerical evaluation.

2.9 IMPORTANT TERMS

The following important terms have been defined in this chapter. You should not go on to Chapter 3 until you thoroughly understand each of them. Each is in italics where it is first defined.

- Optimization under constraints
- Mathematical model
- $n! =$ factorial n
- Duality
- Independent (zeros)
- Alternate optima
- Northwest-corner rule
- Prime
- Chain (of reassignments)
- Closed chain
- Suboptimization
- Systems viewpoint
- Objective function
- Dummy

Problems

2-1 (*a*) Use the assignment algorithm to solve the following problem:

1	0	6	8	2	3	6
1	5	9	1	3	7	6
9	0	3	4	4	9	4
4	3	6	2	4	3	6
9	0	9	1	9	7	6
5	3	6	8	7	8	6
2	5	9	3	0	5	9

(*b*) Find all alternate-optimal solutions (if any). This problem is continued in Problem 7.5-1.

2-2 You are the swimming coach and must designate a men's medley relay team: four swimmers, one of whom will swim free style, one breaststroke, one backstroke, and one butterfly. The objective is to win a race by swimming in a shorter total time than any other similar team. Your team has eight men, each of whom is good (that is, among the best four) in at least one of these strokes, and several are good in two or more of them. You know exactly how good each man is in each stroke; for example, John Smith can swim 100 yards in 49.3, 61.8, and 53.7 seconds in the free style, breast, and butterfly, respectively; he does not swim backstroke. You have similar data on the other seven swimmers. How do you choose the best team? (This problem is continued in Problem 7.4.2-1.)

2-3 Farmer Brown has 15 animals which he would like to sell as quickly as possible because his barn has burned down and his animals are rapidly deteriorating in the harsh environment. For example, the brown heifer, which is worth \$300 if he can get it to market today, is losing value at the rate of 10% per day so that it will be worth only \$270 tomorrow, \$243 the next day, and so on. For each animal, Farmer Brown has exact knowledge of its present market price and its particular rate of deterioration. Unfortunately, he can get only one animal to market per day, because he has only one small truck in which to transport it. How shall he determine in which order to ship these animals to market?

Drill Problems

2-4 Find the unique-optimal solution to the following assignment problem:

16	10	3	12	3	15
1	23	12	5	5	20
21	15	14	23	5	8
37	15	16	14	18	10
3	15	18	25	9	11
10	5	24	15	8	16

This is pretty easy, because you will not have to use step 8 of the algorithm if you follow the algorithm exactly. (Problem 2-9 explores what happens if you change the algorithm by subtracting first from columns.)

2-5 (*a*) Find all minimum-cost solutions to the following assignment problem:

4	0	22	16	13	11
27	0	9	20	16	8
0	0	32	6	14	9
16	8	41	0	0	15
12	28	0	11	8	22
29	13	10	6	17	0

This looks easy, because you can skip most of step 1 (there is already a zero in each row and column), and in step 2 you will immediately find five (of the required six) independent zeros. However, you will have to go through step 8 more than once.

(*b*) Find all maximum-cost solutions (you must go through step 8 five times!).

2-6 Having replaced a small digital computer with a new model, you must modify the off-line automatic electric typewriter to punch a different code. The typewriter punches a six-hole paper tape, and there are thus $2^6 = 64$ possible codes. There are 46 keys on the typewriter, and each punches one of the 64 patterns of hole–no hole on the paper tape. For example, using 1 to represent a hole and 0 to represent no hole, the letter A is 111100 in the code for the old computer and 011010 in the code for the new computer. The typewriter presently punches the former, and must be modified to punch the latter. For the letter A, then, the second, third, and sixth bits are identical, and no changes need be made; the first, fourth, and fifth bits differ, and three corresponding mechanical changes must be made in the steel code bar connected to the A-key; each change involves removing metal in one place and adding metal in another, so that it will activate the proper crossbars. However, these crossbars are connected electrically to the punch, and by simply interchanging the fourth and fifth wires, the number of mechanical changes which need to be made is reduced from three to one. On the other hand, this change of wires may not be good for the 45 symbols other than A. The problem, then, is how optimally to connect the wires from the crossbars to the punch so as to minimize the number of mechanical changes on the code bars.

Some Extra Drill Problems

2	9	2	7	1	
6	8	7	6	1	Cost 13
4	6	5	3	1	
4	2	7	3	1	
5	3	9	5	1	3 solutions

5	5	M	2	6	M is a very large number (maybe 100), which means this assignment is impossible
7	4	2	3	4	Cost 15
9	3	5	M	3	
7	2	6	7	2	
6	5	7	9	1	3 solutions

If you want more practice, make up your own problems, using a table of random numbers. The third-and-fourth-from-last digits of telephone numbers (omitting obvious duplicates) make adequate two-digit random numbers. There are no "bad" problems, and it is unlikely that you will generate one that is exceptionally easy or hard.

Advanced Problems

2-7 (*a*) Prove the statement, in step 7, that "there exists a unique chain." To do this you must prove first that it exists, and second that it is unique. Existence means, for example, that if you have found a starred zero in the chain, there must be at least one primed zero in its row; and uniqueness means in this context that there must be at most one primed zero in its row.

(*b*) Prove that "there exists a unique chain" may not be valid if you attempt to go backward through the chain.

2-8 Write a computer program for solving $n \times n$ assignment problems, where $n \leq 20$. Debug it on our 3×3 examples; then try it on our 10×10. Primes, stars, and coverings will be handled by defining appropriate subscripted variables which take on values of zero or one.

2-9 Problem 2-4 was constructed to be easy if you subtract first from rows and then from columns in step 1. This particular problem is a little harder if you exercise the optional choice of subtracting first from columns. Try it. Note that the final matrix is different, but the optimal solution (the set of independent zeros) remains the same. This possibility of superficial differences in the final matrix is related to the degeneracy of assignment problems as explained in Chapter 6. Save this problem—it is further explored in Problem 6-12.

2-10 (*a*) Make up a "hard" (though small) assignment problem—that is, one that requires going through steps 7 and 8 several times.

(*b*) Is it still hard if one makes optional changes in the algorithm (for example, "star the bottommost zero in the first column"—or the last column; or "cover every row which has a star"; and so forth)?

2-11 Complete the algorithm for finding alternate optima. This is rather tricky, because in scratching out zeros it is not sufficient to go through once looking for rows which have only one zero and then once looking for similar columns. You have to make sure that you have gone through every row and every column without scratching out a new zero before you go on to the next stage of the algorithm. In the next stage, you must write a precise description for finding the closed chains, and you must worry about multiple alternate optima.

2-12 In Section 2.4 it says, "any zeros that remain after the deletion is complete are probably members of alternate-optimal sets." Why "probably?" Specifically,

(*a*) Construct an example in which a zero which cannot be deleted by the given rules is not a members of alternate-optimal sets." Why "probably?" Specifically,

(*b*) Show that for such a zero to exist, there must be at least one other unstarred zero in its row, at least one other unstarred zero in its column, and at least two additional unstarred zeros in two other rows (and columns) of the matrix. Hence there must be several alternate-optimal solutions.

Note: We have also assumed that, after the optimal solution has been obtained, further additions and subtractions would not yield additional zeros. But this is not exactly true, as we explore further in Problem 6-12*b*.

3

THE TRANSPORTATION PROBLEM

We regret to inform you, after you have spent so much time learning about the assignment problem, that this mathematical model is virtually without real practical applications. The difficulty arises in attempting to determine precise values for the c_{ij}'s. If the staff psychologist tells you that the performance of the third worker on the seventh job is 4.72, you would probably do just as well to round it off to 5, because the chances are that the psychologist doesn't know any more about it than you do. But if the numbers in the matrix are themselves not very good, then there isn't much point to a precise optimization; that is, any sort of reasonably good guess at an optimal assignment is probably just as good as the one you'd work out with our algorithm.

In this chapter we introduce a problem which has important practical applications; it is being used routinely by thousands of organizations to minimize their costs. It is called the *transportation problem* and is only a slight modification of the assignment problem. In fact, we shall indicate that the transportation and assignment problems are in some sense identical.

We present in Figure 3-1 a rather special assignment problem in which several of the workers are identical to one another in their performance on all jobs, and in which several of the jobs are also identical. You may ask, "How

3	7	7	7	7	4	4
3	7	7	7	7	4	4
6	2	2	2	2	8	8
6	2	2	2	2	8	8
6	2	2	2	2	8	8
1	9	9	9	9	4	4
1	9	9	9	9	4	4

FIGURE 3-1
An assignment problem.

1	4	2
3	7	4
3	7	4
6	2	8
6	2	8
6	2	8
1	9	4
1	9	4

FIGURE 3-2
Another representation of Figure 3-1.

can two workers be identical? It's not humanistic." And we say, "If the union contract says they're identical, they're identical."

Now clearly a lot of ink has been wasted in setting Figure 3-1. Since the second, third, fourth, and fifth columns are identical, we have to write this column only once; then we can write the number 4 above it to indicate that it should be repeated four times (Figure 3-2). Similarly, the last two columns need be written only once, with the number 2 above. Thus, Figure 3-2 is a condensed form of Figure 3-1 and represents the same problem. We can condense it still further by writing the identical first and second rows only once with the number 2 to the left, and so forth, and this has been done in Figure 3-3. We hope you will convince yourself that Figures 3-1 and 3-3 represent the same problem. That is, Figure 3-3 is a compact notational device for representing any assignment problem (such as Figure 3-1) in which there are several identical rows and several identical columns. Any matrix such as Figure 3-3 represents a so-called transportation problem, and we have proved that a transportation problem is a special case of the assignment problem; namely, it constitutes an assignment problem in which several of the rows and several of the columns are identical.

	1	4	2
2	3	7	4
3	6	2	8
2	1	9	4

FIGURE 3-3
Transportation problem representation of Figure 3-1.

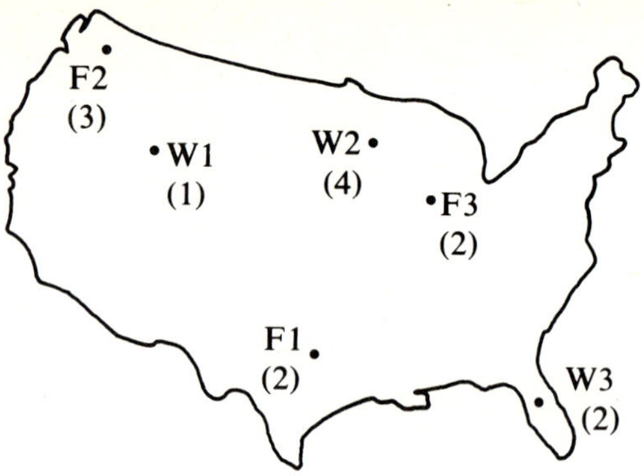

FIGURE 3-4
The transportation problem of Figure 3-3.

The reason why this is called a transportation problem is made clear in Figure 3-4. Assume that the Widget Manufacturing Co. has three factories located as indicated—namely, F1 at Houston, which produces two carloads of widgets a month; F2 at Seattle, which produces three carloads of widgets a month; and F3 at Cleveland, which produces two carloads of widgets a month. It also has three warehouses: W1 located at Denver requires one carload of widgets a month; W2 in Chicago and W3 in Miami require four and two carloads, respectively. The correspondence between Figures 3-4 and 3-3 is further indicated by rewriting Figure 3-3 as Figure 3-5, in which the rows of the matrix are identified with the corresponding factories and the columns of the matrix with the corresponding warehouses. Now we further assume that the cost of sending a carload of widgets from Houston (F1) to Denver (W1) is three units of money (maybe $300?). (The numbers in this example have been arbitrarily chosen as small integers, and we hope you will not attempt to draw any conclusions about actual freight rates from them.) The remaining numbers in the first row represent the cost of shipment of a single carload from Houston to Chicago and from Houston to Miami, respectively.

	W1(1)	W2(4)	W3(2)	
F1(2)	3	7	4	Houston
F2(3)	6	2	8	Seattle
F3(2)	1	9	4	Cleveland
	Denver	Chicago	Miami	

FIGURE 3-5
Expansion of Figure 3-3 to show equivalence with Figure 3-4.

Our problem is to find how to ship our seven carloads of widgets from their origins (the factories) to their destinations (the warehouses) at minimal cost. Of course there are many alternative ways of doing this, and in a large problem it would not be obvious which was the best way.

Since there are two cars full of widgets at Houston, we could, if you like, paint the numbers 1 and 2 on them, and the numbers 3, 4, and 5 on the cars at Seattle, and the numbers 6 and 7 on the cars at Cleveland. Then we could number the loading docks at the destinations: 1 at Denver; 2, 3, 4, and 5 at Chicago; and 6 and 7 at Miami. The problem then becomes one of assigning each of the seven numbered cars to one of the seven numbered loading docks. This argument merely indicates once again the equivalence of Figures 3-1 and 3-3.

We now write the formal matrix representation of the transportation problem. Regretfully, we introduce two new symbols: α, the Greek letter alpha, and β, the Greek letter beta. They are equivalent to the letters *a* and *b*, but we shall use *a* and *b* for other things, and it seems less confusing to ask you to learn these Greek letters than to use *a* and *b* ambiguously.

Given: $\|c_{ij}\| \quad \|\alpha_i\| \quad \|\beta_j\|$

Find: $\|x_{ij}\|$

$$
\begin{aligned}
&\text{Subject to:} && \sum_j x_{ij} = \alpha_i && i = 1, \ldots, m \\
&\text{and} && \sum_i x_{ij} = \beta_j && j = 1, \ldots, n \\
&\text{and} && x_{ij} \geq 0 && i = 1, \ldots, m;\ j = 1, \ldots, n \\
&\text{To minimize:} && z = \sum \sum c_{ij} x_{ij}
\end{aligned}
\tag{3-1}
$$

Compare it with (2-2) on page 56. Here again we are given a matrix of costs—the c_{ij}'s—and asked to find a matrix of x's, each x_{ij} being the number of carloads of widgets shipped from origin i to destination j. This time, however, we are also given a set of α_i, where α_1 is the number of units available at origin 1 (in this case F1), and so forth, and the β's represent the number of units required at the various destinations (thus, β_1 equals 1 and β_2 equals 4.) The first set of constraints says that the total number of units shipped from each origin must equal the total number of units available at that origin; the second set of constraints says the same thing about the destination requirements; the third set of constraints we have written in the linear inequality form (you cannot ship a negative number of carloads), and you remember that we guaranteed (and promised to prove) that when we solve this problem the answers will come out in integers. Finally, we are asked to minimize the total transportation cost.

Now note a surprising fact: If every $\alpha_i = 1$ and every $\beta_j = 1$, then the transportation problem of (3-1) becomes identical to the assignment problem of (2-2). In other words, the assignment problem is a special case of the transportation problem in which the amounts available at each origin and the amounts required at each destination are one. We showed earlier that the transportation problem was a special case of the assignment problem, and we have shown now that the assignment problem is a special case of the transportation problem. It follows that the two problems are identical.

It also follows that since we know how to solve the assignment problem, we already know how to solve the transportation problem. However, in practice we shall wish to make some changes in the algorithm. The reason is that in real transportation problems the numbers α_i and β_j may be quite large, and we don't wish to handle them one row or one column at a time; that is, we wish to make the assignments in chunks. We now show how to do this.

3.1 SOLVING THE TRANSPORTATION PROBLEM

We begin as before by subtracting from each row its smallest element (3, 2, and 1, respectively) and then from each column its smallest element (0, 0, and 1, respectively). The resulting matrix is shown in the upper left of Figure 3-6, and we have indicated the amount subtracted from each row and column because it is most important that we keep track of these. Note that a number such as -3 on the first row indicates that we have really subtracted 3 from each of the first two rows of the corresponding assignment problem. Thus the 0 in (2,2)—you remember that this notation means the element in the second row and second column—really represents 12 zeros, because the second row is really three rows and the second column is really four columns. We do not have to keep track of this in detail, but it means we will have to apply different meanings to our stars, primes, and coverings. For example, if we chose to star some of those 12 zeros, we would first have to decide how many we were going to star; this would be at most three (do you see why?). There are a lot of ways of selecting three independent zeros out of these 12, and it doesn't matter which three we select. Thus, we repeat, stars, primes, and coverings will all have slightly different meanings in the algorithm that we are now developing.

To start, we pick the leftmost column (as usual), and select the first zero, which is in row 1. This element represents a route from origin 1 to destination 1, whose cost in the original matrix was 3, but whose relative cost, indicated in Figure 3-6, is 0. We decide to ship as much as possible along this route; but how much can be shipped? Obviously not more than one carload, because that's all that is required at the first destination. We introduce here a new word, *discrepancy*; it represents the amount yet to be shipped from any origin or the amount yet to be shipped to any destination. The objective of our algorithm will be to reduce all discrepancies to 0. In this case, we assign one unit to this route and reduce the discrepancies of the first row and the first column by one each, as shown in the upper right matrix. We now introduce another new word, *quota*; it represents the amount assigned. In this case, we say that the quota of (1,1) is 1.

Now we note that the 0 in (3,1) cannot be used because the discrepancy of the first column is zero; we therefore move on to the second column and find a zero in (2,2). The amount to be assigned is the smaller of the two discrepancies: 3,which is the row discrepancy, and 4, which is the column discrepancy. Therefore, we assign a quota of 3 to this route and reduce each of the corresponding discrepancies by 3. Since there are no other zeros in the second column, we move on to the third column, where we find a zero in the first row; the most that can be shipped via this route is 1. Corresponding assignment of quota and reduction of discrepancies are performed, and your work should now look like the matrix in the lower right of Figure 3-6 (you can work in pencil on this one or on the first one, or you can copy the numbers on a sheet of scratch paper and work there).

Back in the assignment algorithm, at this point, we covered every column which had a starred zero. However, we are going to use stars differently this time. What we really did was to cover every column where no more assignments could be made, and we do the same thing this time, only our algorithm now reads "cover every column whose discrepancy is 0." In this case, the only such column is the first one, and so you should cover the first column in pencil.

Now we proceed in the familiar fashion. Look for an uncovered zero, find one in (2,2), and prime it. Before, we were asked at this point to look for a star in the same row, which meant check to see whether an assignment can be made to this row. Our corresponding instruction now is to check the discrepancy of the row. If that discrepancy is positive (that is, greater than zero), we have our chain (a new independent zero); if that discrepancy is zero, cover the row (which you should do now) and look for an uncovered zero. Our next uncovered zero occurs in (1,3), and you should prime this zero and cover its row. Your diagram should now look like the last (lower left) matrix in Figure 3-6.

We remind you that, when some of the rows and columns are covered and others are uncovered, every element in the matrix is uncovered or covered once or covered twice. In this last matrix, for example, two elements are uncovered, two elements are twice-covered, and the remaining five elements are once-covered (which ones?). This is the point in our algorithm at which stars are introduced. They will be used as before to establish potential chains of reassignment. They are placed on only those zeros which are candidates for inclusion in such a chain, by virtue of having a quota—that is, by having an assignment which they can give up (you remember that when we go through our chain, we increase assignment at the primed zero and decrease it at the starred zero). Therefore, we cannot put a star on any zero which does not already have a quota. For this purpose we introduce a new name, *essential zero*, which means a zero which has a quota.

Why is such a zero "essential?" We are attempting to find a zero-cost solution to this problem, that is, a solution where shipment is assigned only to those routes whose relative cost is zero. In other words, each product of the form $c_{ij}x_{ij}$ will be equal to zero, either because the c_{ij} is zero (meaning we have

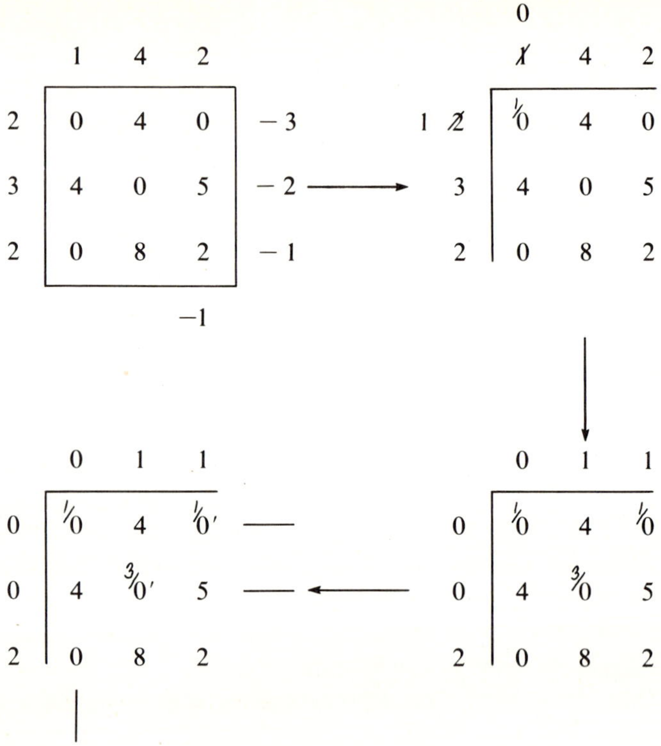

FIGURE 3-6
Solving the transportation problem (start).

developed a zero in the cost matrix) or because the x_{ij} is zero (meaning that there is no quota). Now if a zero has a quota, which means that the x_{ij} is greater than 0, it follows that we must not change the c_{ij} from 0 to something else when we go through the adding and subtracting process. Thus, such a zero is essential. As we shall see, some of the zeros are not essential, namely, those which do not have positive quotas, and occasionally we end up adding to such zeros and thereby losing them.

Now you remember that only the twice-covered elements get added to as the net result of our adding and subtracting. The way, therefore, to ensure that we do not lose our essential zeros is to see that they are not twice-covered. The algorithm does this by uncovering the column of any such zero whenever its row has just been covered. At the same time we star it, indicating that it is a candidate for a chain of reassignments. Thus, the algorithm reads: "For each twice-covered essential zero, star the zero and uncover its column." In this case, the zero in (1,1) is essential and twice-covered, so we star it and uncover the first column.

The next step is routine. Having just uncovered the first column, we check in this column for an uncovered zero, find one in (3,1), prime it (Figure 3-7), and check the discrepancy of the row. We note that the discrepancy is

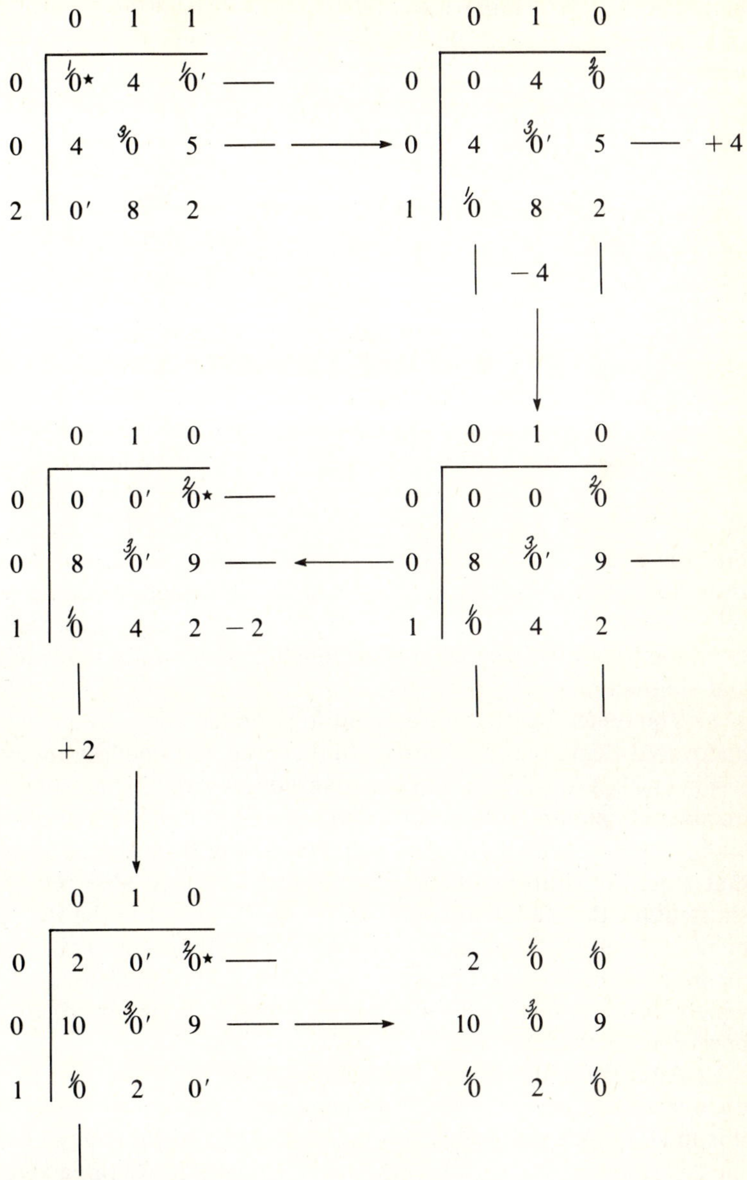

FIGURE 3-7
Solving the transportation problem (conclusion).

positive, and therefore we have found a method of making new assignments at zero cost (we would earlier have said that we found a new independent zero). Specifically, our chain goes from (3,1) to (1,1) to (1,3); however, going through this chain becomes a little more complicated, because we would like to increase and decrease the quotas through this chain by as much as possible. Note that in going through the chain we will add to the quotas of the primed zeros and subtract from the quotas of the starred zeros. We will also subtract from the discrepancy of the third row (the row where the chain starts) and from the discrepancy of the third column (the column where the chain ends). Each of these subtractions places a constraint on the amount; that is, we cannot subtract more than either of these discrepancies, nor can we subtract more than the quota of any starred zero in the chain. The amount to be added and subtracted in going through this chain is therefore the smallest of these amounts; in this case it is 1. Therefore we go through the chain, subtracting 1 from the discrepancy of the third row, adding 1 to the quota of the primed zero in (3,1), subtracting 1 from the quota of the starred zero in (1,1), adding 1 to the quota of the primed zero in (1,3), and subtracting 1 from the discrepancy of column three. All these things have been done in the upper right matrix of Figure 3-7. Note that the nonessential zero in (3,1) has become essential, and the essential zero in (1,1) has become nonessential; it turns out that it will later be lost.

Now erase all primes, stars, and row coverings, and go back to the beginning of the algorithm. Cover every column whose discrepancy is zero; this time there are two such columns. Proceed as before: Look for an uncovered zero, find it in (2,2), prime it, and check the discrepancy of the row; finding it is zero, we cover the row. Now we look for an uncovered zero, find none (second matrix in Figure 3-7), and therefore conclude that we are ready to do the adding and subtracting.

You remember that the amount to be added and subtracted is the smallest uncovered element in the matrix. In this case, the smallest uncovered element is 4 in (1,2). We will add 4 to the covered row (row 2) and subtract 4 from the uncovered column (column 2). You remember that we had the alternative of adding to the covered columns and subtracting from the uncovered rows, but that would obviously be harder in this case. You also remember that our shortcut for the actual computation is simply to add to the twice-covered elements (of which there are two here) and subtract from the uncovered elements (of which there are also two here). The once-covered elements (of which there are five) will remain unchanged. The appropriate arithmetic has been done in the third matrix of Figure 3-7.

After the addition and subtraction, there will always be at least one new uncovered zero where the smallest uncovered element was (why?). In this case it is in (1,2). We prime this zero and check the discrepancy of the row; since the discrepancy is 0, we cover the row. Now there are three zeros in this row, but the one in the first column is not essential, and the one in the second column is neither twice-covered nor essential. However, the one in the third column is both, and we star it and uncover the third column (fourth matrix in Figure 3-7). We now note once more that all zeros are covered, and we have discovered a new (and unpleasant) phenomenon: After we have added and subtracted and generated new zeros, we do not necessarily get new independent zeros, or, if you like, we do not necessarily get a new or improved assignment. This may

	1	4	2	
2	3	$\overset{1}{7}$	$\overset{1}{4}$	3
3	6	$\overset{3}{2}$	8	−2
2	$\overset{1}{1}$	9	$\overset{1}{4}$	3
	−2	4	1	

FIGURE 3-8
The complete solution.

also happen in the assignment problem (although it did not happen in our example). On the other hand, we can get lucky: We might get to the chain step two or more times before we have to do more adding and subtracting.

In this case, we must now add and subtract again. The smallest uncovered element is 2 in (3,3), so we choose this time to add 2 to the covered column (the first column) and subtract 2 from the uncovered row (the third row). The arithmetic is shown in the fifth matrix. Our new uncovered zero in (3,3) is now primed, and we note that the discrepancy of the row is greater than 0, so we have a chain: from (3,3) to (1,3) to (1,2) to the nonzero discrepancy of the second column. Reassigning through this chain, we arrive at the last matrix in Figure 3-7. All discrepancies are now 0, so we are finished. All primes, stars, and coverings may be removed, as may the zero discrepancies.

We now transfer the solution (that is, the quotas obtained) to the original problem, copied from Figure 3-4, to yield Figure 3-8. We first check discrepancies against quotas; for example, the discrepancy of 2 in the first row must equal the total of all the quotas in the first row (which, of course, it does), and similarly for the other rows and columns. In Figure 3-8 we have also cumulated (to the right and below the matrix) the total net amounts subtracted from each row and column. Note, for example, in Figure 3-6 that we subtracted 3 from the first row, and so 3 has been written to the right of the first row in Figure 3-8 (pay special attention to the fact that we have changed the sign here, because the amounts we are interested in are the amounts that have been subtracted). In row 2 we subtracted 2 in Figure 3-6, but we added 4 in Figure 3-7; therefore the net result is to add 2 to this row, or to subtract −2; and −2 has been written to the right of the second row in Figure 3-8. Check the other numbers which indicate the total net amounts subtracted from rows and columns.

We must now compute the cost which we have found (the minimum transportation cost for this problem), and as before we shall do it two ways. First note that 3 has been subtracted from the first row, but there are really two of these rows, as indicated by the number 2 on the left of the matrix, and so the total amount we have subtracted here is 6. Similarly, we have subtracted −6 from the second row, and we have subtracted 6 from the third row. We have also subtracted −2 from the first column, 16 from the second column, and 2 from the third column. If we add up all these numbers, we find that we have subtracted 22 in going from the original problem (Figure 3-4) to the final solution (Figure 3-8). The cost may also be computed directly from Figure 3-8 by multiplying each cost by its quota and adding, and this number, of course, also comes to 22.

It is most satisfying to see this equivalence, but it is more than that; it is a check on the accuracy of our work. We have already performed two checks (that the final quotas add up to the original discrepancies and that the costs computed in two ways are equal). One more check is possible. If we had been terribly smart, instead of subtracting 2 from the second row at the beginning of Figure 3-6, we would have added 2 to this row, since it turns out that this is eventually what we had to do. Of course, there's no way that anybody can be that clever, but now that we know the answer, we can check it. Thus, if we take the cost numbers in Figure 3-8 and subtract the appropriate amounts (subtract 3 from row 1, subtract −2 from row 2, and so forth), we will get precisely the numbers shown in the lower right matrix of Figure 3-7. If all three of these checks work out, you can be quite confident that the final solution is correct. We now present the formal algorithm.

3.2 ALGORITHM FOR THE TRANSPORTATION PROBLEM

Step 1 Subtract from every row its smallest element. Subtract from every column its smallest element. Keep track of the amounts subtracted. Locate the topmost zero in the first column. Go to step 2.

Step 2 Note the discrepancies of the row and column of this zero. The smaller of these numbers (or, if they are equal, this number) is assigned as a quota to this zero and is subtracted from both discrepancies (so that at least one discrepancy becomes zero). If the column discrepancy is now 0, go to step 3; if it is positive, go to step 4.

Step 3 If this is the last column, go to step 5. If not, choose the next column, find the topmost zero, and go to step 2.

Step 4 Look downward in the column for another zero in a row with a positive discrepancy. If one is found, go to step 2; if none is found, go to step 3.

Step 5 Cover each column whose discrepancy is 0. If all columns are covered, go to step 11; if there are any uncovered columns, go to step 6.

Step 6 Look for an uncovered zero. If one is found, go to step 9; if none is found, go to step 8.

Step 7 There now exists a unique chain (which may contain just one zero, or three, or five, ...) as follows: Go vertically from the uncovered primed zero to the starred zero in this column, if any. From there go horizontally to a primed zero. From there go vertically to the starred zero in this column, if any. And so forth. The chain ends when you arrive at a primed zero which has no starred zero in its column. The discrepancy of this column will be positive. Check for the smallest of the following numbers: the discrepancy of the row of the first primed zero in the chain (the uncovered primed zero), the quota of every starred zero in the chain, the discrepancy of the column of the last primed zero in the chain. This number is to be subtracted from each of these two discrepancies, and from the quota of every starred zero in the chain, and added to the quota of every primed zero in the chain. Now erase all primes, stars, and row coverings. Go to step 5.

Step 8 Find the smallest uncovered element in the matrix; add this to the covered rows and subtract it from the uncovered columns (or add it to the covered columns and subtract it from the uncovered rows, or add it to the twice-covered elements and subtract it from the uncovered elements). Keep track of the amounts added and subtracted. Do not change any primes, stars, or coverings. Go to step 6.

Step 9 Prime this zero. Check the discrepancy of this row. If it is zero, go to step 10; if it is positive, go to step 7.

Step 10 Cover this row, and check all the zeros in this row. For each twice-covered essential zero, star the zero and uncover its column. Go to step 6.

Step 11 Uncover all columns, cumulate the amounts subtracted from rows and columns, and write these net totals to the right and below the original matrix. Transfer the final quotas to the original matrix. Perform three checks as follows:

(*a*) For each row and column, the sum of the quotas should equal the discrepancy.

(*b*) By subtracting the indicated amounts from rows and columns it should be possible to go directly from the first matrix to the final matrix.

(*c*) The cost computed in two different ways should be equal: First, multiply each cost by its quota and add them; second, multiply each original discrepancy (on the left or above) by the corresponding amount subtracted (on the right or below) and take the algebraic sum of these products. If alternate-optimal solutions are desired, go to step 12; otherwise stop.

Step 12 Left as an exercise; alternate optima are discussed below.

We stated "algebraic sum" in step 11*c* to remind you that some of these products which will be added are positive and others may be negative. Thus, if you are adding +7 and −2, the algebraic sum is 5, not 9 (the "sum" is also 5, but we wanted to be sure that you would remember). We have chosen as an example a small problem, but it has exercised almost every aspect of the algorithm that would have been exercised by a much larger problem. There are only two exceptions: In large problems there may be more than one twice-covered essential zero in the row, and, in this case, our algorithm instructs us to star each of them and uncover its column. The other point is that in large problems, the amount that can be transmitted through the chain will often be greater than 1 (it happened to be only 1 in our example), and it may be limited either by the smaller of the two discrepancies in the chain or by the smallest quota of any starred zero in the chain. When you go through the chain, adding and subtracting, at least one of these quotas or discrepancies will go to zero.

The hints for doing this algorithm are similar to those we had before. Take your time. Write all the costs in ink; discrepancies, quotas, stars, primes, and coverings should be in pencil. When a quota becomes 0 (or if you are instructed to write a zero quota in step 2), erase it. When you come to step 8 (adding and subtracting), write a new cost matrix. You should note at this point which rows and columns are being added to and subtracted from, and write these additions and subtractions next to the first matrix. When the problem has been completed, cumulate these numbers (don't forget to change their signs). If at any time you get confused, erase all pencil marks on the last matrix, cumulate the amounts added and subtracted, and make sure you can go directly from the first matrix to the last one you now have. If this last matrix is incorrect, go back until you find one that is correct. In this last correct matrix, erase all pencil lines, put back the original discrepancies, and start with step 2 of the algorithm. At the end of step 7, when you've gone through a chain, it might be a good idea to perform the first check; that is, make sure that, row by row and column by column, the quotas plus the remaining discrepancy add up to the original discrepancy.

3.3 ALTERNATE OPTIMA

As with everything else about the transportation problem, alternate optima are a little more complicated than in the assignment problem. The solution of our sample problem was unique, but this need not be the case. Alternate optima exist whenever we can find closed chains of zeros in the final cost matrix with essential zeros at every even-numbered position in the chain; that is, the second, fourth, and so on, zeros must be essential. You can start counting the chain at any point. The total number of zeros in the closed chain must be even. We can then go through this chain, subtracting from the quotas of the even-numbered zeros and adding to the quotas of the odd-numbered zeros, to develop an alternate-optimal solution. The amount to be added or subtracted may

	5	12	2	6
11	3	11/0	27	0
6	5/0	1/0	8	44
8	0	19	2/0	6/0

(*a*)

	5	12	2	6
11	3	6/0	27	5/0
6	0	6/0	8	44
8	5/0	19	2/0	1/0

(*b*)

FIGURE 3-9
Two alternate-optimal basic solutions to a transportation problem. (*a*) Degenerate; (*b*) nondegenerate.

be any number from 1 up to the smallest quota of the even-numbered zeros in the chain.

Suppose, for example, that our final solution looked like the matrix on the left of Figure 3-9. We note in passing that this matrix is not square: it is 3×4. But it represents a square (25×25) assignment problem, so it is properly balanced. That is,

$$\sum_i \alpha_i = \sum_j \beta_j \qquad (3\text{-}2)$$

If these two sums were not equal, we would have to add dummies to balance it, introducing the appropriate costs, as before. If, for example, the demand (at the destinations) exceeded the supply (at the origins), we should have to add a dummy row in which the cost for each destination was the penalty for shortage there.

In the matrix on the left of Figure 3-9, there does exist a closed chain, starting at (1,4) and going successively to (3,4), (3,1), (2,1), (2,2), (1,2), and back to (1,4). There is an even number (six) of zeros in this chain; if we arbitrarily call the one in (1,4) the first, we notice that the even-numbered ones—namely, the second, fourth, and sixth—are all essential and therefore that alternate-optimal solutions do exist. We may go through this chain, subtracting any number up to a maximum of five (why?) from the even-numbered zeros in the chain and adding the same number to the odd-numbered zeros. Note, therefore, that there are six different optimal solutions to this problem, and note how each of them has the same cost and satisfies all the original discrepancies. We have indicated one of these on the right of Figure 3-9, where we have added and subtracted 5. The two alternate-optimal solutions that are shown in the figure are "basic solutions," and we shall later show that they are indeed basic solutions in precisely the sense which we have defined in Section 1.4. It is a general principle of linear programming, as we shall see, that when we have found alternate-optimal basic solutions, all intermediates between ("nonbasic solutions") are also optimal; we shall investigate these in Problem 3-7. We shall return to this point, and we shall also return to these figures to indicate something about degeneracy, but all that requires more knowledge of linear programming than we have at the moment.

3.4 PRIMAL AND DUAL ALGORITHMS

We have presented above, as equations (3-1), the formulation of the transportation problem analogous to the one that we presented for the assignment problem in (2-2). We also presented in (2-3) a formulation of what we called the *dual assignment problem*. The transportation problem has an analogous dual, in which at appropriate points we substitute α_i or β_j for 1. This formulation is

Given:	$\|c_{ij}\|$ $\|\alpha_i\|$ $\|\beta_j\|$	
Find:	$\|u_i\|$ $\|v_j\|$	
Subject to:	$u_i + v_j \leq c_{ij} \qquad i = 1, \ldots, m;\ j = 1, \ldots, n$	(3-3)
To maximize:	$w = \sum_i \alpha_i u_i + \sum_j \beta_j v_j$	

You remember that the u_i and v_j are the amounts subtracted from rows and columns, respectively. Because this is a dual problem, the variables of the problem, that is, the u_i and v_j, are called *dual variables*; the optimal values of these variables are written to the right and below the final matrix (Figure 3-8). Thus, the optimal value of u_1 is 3, of u_2 is -2, and so forth. The sums which we are asked to maximize are called *weighted sums*, in which each u_i is weighted by (that is, multiplied by) its corresponding α_i, and each v_j by β_j. We have already computed this sum of weighted sums; the total was 22 in our example and represented the total cost of the problem. It was, of course, equal to the corresponding primal cost, represented by the weighted sum of the x_{ij} (the weights being the c_{ij}); the x_{ij} are the variables of the primal problem and are therefore called *primal variables*.

The algorithm which we developed was primarily devoted to manipulating the dual variables in such a way as to maximize the weighted sum of dual variables, and so it was a *dual algorithm*; it determined for us incidentally the optimal values of the primal variables. However, we could have developed a *primal algorithm*—that is, an algorithm in which we ignored the dual variables completely and instead straightforwardly attempted to minimize the weighted sum of the primal variables, that is, $\Sigma\Sigma\, c_{ij}x_{ij}$. We now show such an algorithm; however, it is plagued with certain difficulties due to degeneracy, about which we will learn more later, and therefore we cannot develop it completely at this time.

In the upper left of Figure 3-10 we have copied our original example problem; we will not do any adding to and subtracting from the costs; we will merely manipulate quotas. Starting, as always, in the northwest corner, we as-

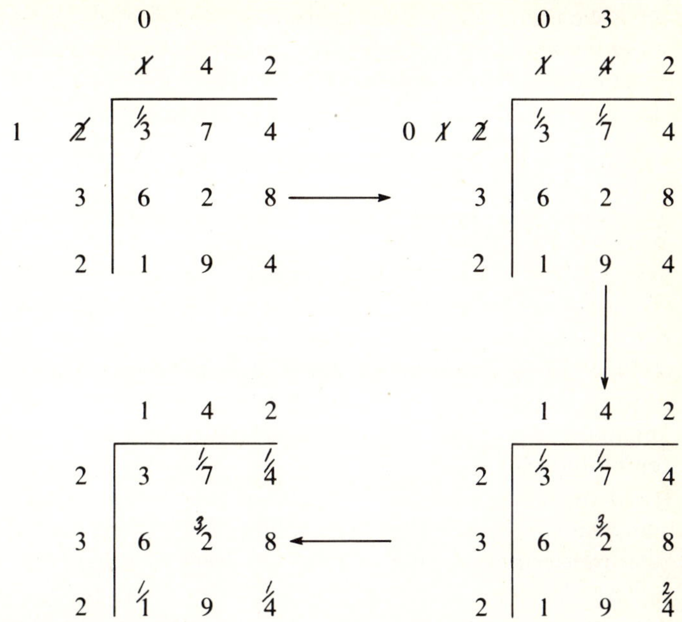

FIGURE 3-10
Solving by the primal algorithm.

sign as much as possible to (1,1)—in this case a quota of 1. Now we have used up the discrepancy of the first column, so we move to the second column and find that the maximum quota we can assign to (1,2) is also 1; this is assigned in the second matrix. Continuing the same process, we assign quotas of 3 to (2,2) and 2 to (3,3) in the third matrix. All the discrepancies have now been reduced to 0 and we need not worry about them; since all discrepancies are always immediately reduced to 0 in this algorithm, we do not need to write these zeros.

We shall now start to see if we can reduce the costs. An obvious place to look is in the lowest cost element of the matrix, namely, the 1 in (3,1). The question is whether this can be a member of a closed chain such as we have discussed above. In fact it can; the chain runs (3,1), (1,1), (1,3), (3,3), and back to (3,1). If we now start at the right place, for example, at (3,1), we can go through this closed chain, reassigning one unit and reducing the total cost. That is, we will add to the quotas of (1,3) and (3,1), which have a total cost of 5, and subtract from the quotas of (1,1) and (3,3), which have a total cost of 7. Thus we will reduce our total cost by 2, which is good. These changes have been shown in the final matrix (lower left of Figure 3-10). We now look for further closed chains in this matrix, but we do not find any which are profitable; that is, there is no closed chain in this last matrix in which we can reassign in such a way as to reduce the total cost. Therefore, this solution is optimal, and it is of course the same solution that we had arrived at previously.

We think you will recognize that we have left this a little vague in spots. After we know enough to make things specific, we will return to this algorithm (Section 6.5). In this particular example, the primal algorithm was easier than the dual algorithm, but as you will see, it is not always so.

3.5 APPLICATIONS

Imagine that you are the vice president of a large petroleum company and that this week you must ship 15 shiploads of oil from Kuwait, 11 from Bahrein, 13 from Indonesia, 6 from Houston, 2 from Los Angeles, ... Of these, 5 must go to New York, 2 to Buenos Aires, 8 to London, 13 to Tokyo, ... The total problem, of which we have indicated only a piece, involves the shipment of 237 shiploads of petroleum located at 16 different origins, to be shipped to 24 different destinations. If you are a very good vice president for transportation, worth the large salary you are getting, you may be able to come up with a solution which is within 1% of the optimum. But that extra 1% is worth saving if your transportation bill for those hundreds of ships is measured in millions of dollars. You would probably think it was worth a thousand dollars of programming and computer time to find the true optimum (or all the alternate optima if there were any). There is another important point: Even if you are both bright and lucky, and happen to hit on the optimum, you couldn't be sure of it. You would lie awake nights worrying about the possibility that by shuffling things around you might be able to save many thousands of dollars. With a computerized solution there is no such problem. Assuming only that the costs are correct, the computer will unerringly indicate the optimum or optima.

This kind of problem is indeed common, and once the computer program has been written, it is very simple to operate. Even a fairly large problem, say, 50 origins, 100 destinations, and 10,000 units to be shipped (a 50×100 transportation problem corresponding to a $10{,}000 \times 10{,}000$ assignment problem) can be solved by a large computer in a few seconds. We warn you, however, that it takes a great deal more than a few seconds to find the 5,000 cost numbers that go into a 50×100 problem.

Fortunately, in realistic cases, most of these numbers would not have to be determined: Where it seemed reasonably obvious that the cost of a particular route was prohibitively large, a rough guess at it might be made. For example, it seems unlikely that you would ship from Houston to Tokyo if there were a convenient destination (New York) close to Houston and a convenient origin (Indonesia) close to Tokyo. But large problems can be confusing, and after the problem had been solved, it might turn out that the computer had selected this expensive route; then it would be necessary to find that cost accurately and solve the problem again. But we are fortunate: It will not really be necessary to re-solve the whole problem, because after we have more understanding of the nature of these problems and their solutions, we will learn that it is possible to find, without very much work, a new solution when something has changed. This also comes later.

In addition to straightforward problems of transporting goods from origins to destinations, which as we have noted are ubiquitous, there are some interesting problems which we can fit into the mathematical model called the transportation problem if we use sufficient imagination. For example, it turns out to be possible to formulate certain kinds of scheduling problems as problems of transporting requirements from one period of time to another, with appropriate penalties (that is, costs) when something is delayed; see Problem 3-3—and remember that this is a problem at the end of this chapter, while the problem (3-1) referred to in the next section is a set of equations at the beginning of the chapter.

Bear in mind the distinction between "formulating" a problem and "solving" a problem. To formulate something like the scheduling problem as a transportation problem requires that one first decide the nature of the origins and of the destinations and then specify the following things: m, the number of rows; n, the number of columns; α_i, the number of units available at the ith origin, for $i = 1, \ldots, m$; β_j, the number of units required at the jth destination, for $j = 1, \ldots, n$; and c_{ij}, the cost of transporting one unit from the ith origin to the jth destination for all i and j. If all these have been specified, and $\Sigma\alpha_i = \Sigma\beta_j$, then we have formulated as a transportation problem and may begin the solution process. The advantage of so formulating is that we happen to have a neat algorithm for solving such problems.

3.6 ANOTHER FORMULATION

The key step in formulating a problem is to define a pertinent set of variables (or unknowns). In this case we have chosen as variables a set of x_{ij}'s, each of which represents the amount shipped from the ith origin to the jth destination. We may choose any values of the variables which satisfy each constraint; and we would like to choose values of the variables which yield the smallest possible value of the objective function. This is the nature of the problem (3-1) which says, "find a set of x's which . . . ." We then met another problem, (3-3), which said, "find a set of u's and v's which . . . ." Note that it has a different set of variables. To distinguish them, we called the first problem (and its variables) "primal" and the second problem (and its variables) "dual." Examining the dual problem helped us to solve the primal; in fact, it turned out that we solved both at once, and both had the same optimal objective function.

In describing the transportation problem of Figure 3-3, and the assignment problem of Figure 2-1, we used the notation shown in Figure 3-11, which is called *double-subscript notation*. That is, each cost and each variable is iden-

c_{11}	c_{12}	c_{13}	x_{11}	x_{12}	x_{13}
c_{21}	c_{22}	c_{23}	x_{21}	x_{22}	x_{23}
c_{31}	c_{32}	c_{33}	x_{31}	x_{32}	x_{33}
	(*a*)			(*b*)	

FIGURE 3-11
Double-subscript notation for a 3×3 problem. (*a*) Cost matrix; (*b*) primal-variables matrix.

tified by two subscripts, the first of which refers to a row (origin or worker) and the second to a column (destination or job). While this notation was particularly convenient for developing our algorithm, a slightly different notation will be more useful in relating these problems to the general linear programming problem which we will study next. This new notation puts only a single subscript on each variable so that the costs (in a 3×3 problem, for example) run from c_1 to c_9 instead of c_{11} to c_{33}, and similarly for the x's (Figure 3-12).

In terms of this single-subscript notation, we now write out in detail all the equations for this problem, which we merely summarized symbolically before. For example, there are three constraints on the rows; the first of these in our old notation said

$$x_{11} + x_{12} + x_{13} = 2$$

which describes the particular problem of Figure 3-3. In our new notation this will say

$$x_1 + x_2 + x_3 = 2$$

In general, the first row constraint on any 3×3 transportation problem will be

$$x_1 + x_2 + x_3 = \alpha_1$$

or, in detached-coefficient form,

$$1 \quad 1 \quad 1 \qquad \alpha_1$$

There are two other similar row constraints and three column constraints. Finally, our objective is to minimize the total cost; that is,

Minimize: $$z = \sum_j c_j x_j$$

We now write out all these equations:

$$\begin{array}{lllllllllr}
x_1 + x_2 + x_3 & & & & & & & & = \alpha_1 & \\
& & & x_4 + x_5 + x_6 & & & & & = \alpha_2 & \\
& & & & & & x_7 + x_8 + x_9 & & = \alpha_3 & \\
x_1 & & & + x_4 & & & + x_7 & & = \beta_1 & (3\text{-}4) \\
& x_2 & & & + x_5 & & & + x_8 & = \beta_2 & \\
& & x_3 & & & + x_6 & & + x_9 & = \beta_3 &
\end{array}$$

c_1	c_2	c_3	x_1	x_2	x_3
c_4	c_5	c_6	x_4	x_5	x_6
c_7	c_8	c_9	x_7	x_8	x_9
	(*a*)			(*b*)	

FIGURE 3-12
Single-subscript notation for a 3×3 problem. (*a*) Cost matrix; (*b*) primal-variables matrix.

Minimize: $c_1x_1 + c_2x_2 + c_3x_3 + c_4x_4 + c_5x_5 + c_6x_6 + c_7x_7 + c_8x_8 + c_9x_9 = z$

or, detaching coefficients,

$$\begin{array}{ccccccccc} x_1 & x_2 & x_3 & x_4 & x_5 & x_6 & x_7 & x_8 & x_9 \\ 1 & 1 & 1 & 0 & 0 & 0 & 0 & 0 & 0 = \alpha_1 \\ 0 & 0 & 0 & 1 & 1 & 1 & 0 & 0 & 0 = \alpha_2 \\ 0 & 0 & 0 & 0 & 0 & 0 & 1 & 1 & 1 = \alpha_3 \\ 1 & 0 & 0 & 1 & 0 & 0 & 1 & 0 & 0 = \beta_1 \\ 0 & 1 & 0 & 0 & 1 & 0 & 0 & 1 & 0 = \beta_2 \\ 0 & 0 & 1 & 0 & 0 & 1 & 0 & 0 & 1 = \beta_3 \end{array} \qquad (3\text{-}5)$$

Minimize: $\quad c_1 \quad c_2 \quad c_3 \quad c_4 \quad c_5 \quad c_6 \quad c_7 \quad c_8 \quad c_9 = z$

We remind you of how far you have come. The arrangement of numbers in (3-5), which would have been complete gibberish to you a little while ago, now represents the problem of minimizing the costs of transporting α_1 units from the first origin, and so forth, where c_1 is the cost of transporting one unit from the first origin to the first destination, and so forth. If all the α's and β's are changed to 1s, this becomes the assignment problem.

Equations (3-5) [or (3-4), which are the same, written differently] are the third matrix representation of the transportation problem which we have seen. These three matrices are given different names: the "cost matrix" of Figure 3-11*a* or 3-12*a*, the "primal-variables matrix" of Figure 3-11*b* or 3-12*b*, and the "constraint matrix" of equations (3-5). We shall meet all of them again, so if one refers to "the transportation-problem matrix," it is important to be sure which matrix is meant.

We have mentioned "duality," which has to do with the interchangeability of rows and columns, and which we said was very important in linear programming. What happens if we interchange the rows and columns in the problem (3-5)? For this purpose, the x's at the top are considered to be labels, not parts of the problem, and the first row of the dual problem [which was the first column of (3-5), the primal problem] becomes

$$1 \quad 0 \quad 0 \quad 1 \quad 0 \quad 0 \qquad c_1$$

If we do the same with every column, the dual problem becomes

$$\begin{array}{cccccccl}
& u_1 & u_2 & u_3 & v_1 & v_2 & v_3 & \\
& 1 & 0 & 0 & 1 & 0 & 0 & \le c_1 \\
& 1 & 0 & 0 & 0 & 1 & 0 & \le c_2 \\
& 1 & 0 & 0 & 0 & 0 & 1 & \le c_3 \\
& 0 & 1 & 0 & 1 & 0 & 0 & \le c_4 \\
& 0 & 1 & 0 & 0 & 1 & 0 & \le c_5 \\
& 0 & 1 & 0 & 0 & 0 & 1 & \le c_6 \\
& 0 & 0 & 1 & 1 & 0 & 0 & \le c_7 \\
& 0 & 0 & 1 & 0 & 1 & 0 & \le c_8 \\
& 0 & 0 & 1 & 0 & 0 & 1 & \le c_9 \\
\text{Maximize:} & \alpha_1 & \alpha_2 & \alpha_3 & \beta_1 & \beta_2 & \beta_3 & = w
\end{array} \qquad (3\text{-}6)$$

The primal problem, (3-5), had six constraint equations (that is, six rows) and nine variables (that is, nine columns), and we called these variables x's. The dual problem has nine rows (and therefore nine constraints) and only six columns (and therefore six variables). We have chosen to call these variables u_1, u_2, u_3, v_1, v_2, and v_3. The first constraint equation of the primal problem said that exactly two units had to be shipped from the first origin. The first constraint equation of the dual problem says that the amount subtracted from the first row (namely, u_1) plus the amount subtracted from the first column (namely, v_1) cannot exceed the original cost in the first row and first column (namely, c_1). And, subject to nine constraints of this nature, one for each element of the cost matrix, we are to maximize w, which is the sum $\alpha_1 u_1 + \alpha_2 u_2 + \cdots$.

Again we have left things a little vague; we have not explained, for example, why the constraints in (3-6) are in the form of inequalities, and we have left out certain related considerations about nonnegativity. We will explain all this in detail later, and at that time we will assert something called *the fundamental theorem of linear programming*, which says, in essence, that when we solve the primal problem and thereby find the minimum value of z, and when we solve the dual problem and thereby find the maximum value of w, these two values will be equal. In this particular case we have already seen the equality of these quantities (both were 22 for our example transportation problem).

In the problem (3-5), if the vector of numbers on the right consists entirely of 1s, then we have a special case called the assignment problem. If they may take on arbitrary values (here we are restricting them to positive integers), it becomes a more general problem called the transportation problem. In either case, the row vector at the bottom of the problem may be any arbitrary numbers. However, the rectangular matrix is highly specialized in both cases, consisting exclusively of ones and zeros, with these ones and zeros in a special symmetric pattern. The transportation and assignment problems are therefore special cases of a generalized problem in which arbitrary numbers may be placed anywhere in this rectangular matrix. This more general problem is called the linear programming problem, and we shall meet it next.

3.7 IMPORTANT TERMS

The following important terms have been defined in this chapter. You should not go on to Chapter 4 until you thoroughly understand each of them. Each is in italics where it is first defined.

- Discrepancy
- Quota
- Essential zero
- Dual variable
- Primal variable
- Dual algorithm
- Primal algorithm

Problems

You may want to solve the very easy Problem 3-4 before trying the more typical Problem 3-1.

3-1 (*a*) Solve the following transportation problem:

	7	5	3	2
6	2	3	11	7
1	1	0	6	1
10	5	8	15	9

(*b*) Find all alternate optima, if any.

This problem is continued in Problem 3-9.

3-2 (*a*) Suppose you had a transportation problem with given α_i (origin requirements), β_j (destination requirements), and c_{ij} (transportation costs) and were asked to find a set of x_{ij} (amounts shipped) to minimize $z = \Sigma\Sigma c_{ij}x_{ij}$ according to the usual constraints of the problem (3-1). Suppose you had found this minimum z—call it z_0. Now change the problem: All the costs remain the same; all the new requirements are at least as large as, and some are larger than, the old requirements. The set of x's which solves this new problem may be different; in any case, call the minimum cost of this new problem z_1. Now the question is, will z_1 necessarily be larger than z_0, or do you think the strongest statement you could make would be $z_1 \geq z_0$?

(*b*) Solve the following transportation problem:

	2	1
2	3	1
1	1	3

(*c*) Solve the following transportation problem:

	2	2
2	3	1
2	1	3

(*d*) Construct a geographic example by choosing four American cities, which makes these example problems reasonable.

Note: This is an important problem and will be referred to subsequently in the text.

3-3 As production manager of the Widget Manufacturing Company, you must schedule production for the next 12 months. The demand, d_k, in the kth month is known exactly and must be met. However, you may choose to produce more in some months than is required and store it to meet later demand, either because production costs may be lower in the earlier month or because production capacity in certain months is insufficient to meet the demands for those months. Production capacity in the kth month is a_k during regular time, at a cost of e_k dollars per widget, and b_k during overtime, at a cost of f_k dollars per widget (where $f_k > e_k$). Inventory costs for widgets on hand at the end of the kth month (that is, stored to meet later demand) are c_k dollars per widget for the month. All 72 of these numbers (a_k, b_k, c_k, d_k, e_k, and f_k) are known. Overall production capacity is adequate; that is, $\Sigma(a_k + b_k) > \Sigma d_k$. Formulate as a transportation problem. You may answer this problem by showing the α_i, β_j, and c_{ij} (as in Figure 3-3) for at least the first several rows and columns or by defining m, n, α_i, β_j, and c_{ij} in terms of the a's, b's, and so forth. The latter, which might be preferred by a mathematician, will probably be much more difficult for you. *Hint*: To ensure that your formulation does not permit nonsense solutions (such as producing in one month to meet demand in a previous month), insert very large fictitious costs (represented by M to symbolize a very large number) in appropriate places.
Note: This is the first time we have asked you to write formulas involving symbols and subscripts and such. It is important that you get practice in doing this sort of thing. If you like, you may first formulate a three-month problem, using the following specific numbers in place of symbols:

	k = 1	2	3
a_k	10	15	10
b_k	15	10	10
c_k	5	4	
d_k	14	3	27
e_k	8	20	6
f_k	12	30	9

Here, for example, in the first month you can produce up to 10 widgets in regular time at \$8 each and up to 15 in overtime at \$12 each. You must make at least 14 for sale; if you make more, you will pay \$5 each for the first month's storage. After you have formulated this smaller, specific problem (or before if you like), you must formulate the general symbolic problem. See also the hint at the end of Section 3.5.

Drill Problems

3-4 Solve the following transportation problem and check it:

	3	4	5
6	0	7	5
2	4	0	3
4	1	6	0

3-5 (*a*) Solve and check. Your solution should show the optimal values of all primal and dual variables.

	3	4	5	6	7
8	3	7	21	4	9
9	81	11	2	18	29
8	6	0	5	37	12

(*b*) Find as many alternate-optimal solutions as you can. This problem is continued in Problem 6-18.

3-6 Army Ordnance needs new lanyards for its various cannons. In particular, three arsenals need this item, as follows:

Raritan Arsenal, Metuchen, N.J.	1000 lanyards
Redstone Arsenal, Huntsville, Ala.	2600 lanyards
Rock Island Arsenal, Rock Island, Ill.	2000 lanyarrds

The following bids are received:

A.C. Parksplug Co., Boston, Mass.	2000 lanyards at $10/100
Pullco, Inc., Detroit, Mich.	2200 lanyards at $8/100
Neverbreak Lanyard Co., Pittsburgh, Penn.	3000 lanyards at $8/100
Peoples' Lanyards, Havana, Cuba	500 lanyards at $6/100

Federal and state regulations (and perhaps graft) cause additional costs per hundred to be incurred, as follows:

From/to →	Ala.	Cuba	Ill.	Mass.	Mich.	N.J.	Penn.
Alabama	0	4	14	12	12	9	8
Cuba	4	0	19	17	18	14	13
Illinois	14	19	0	13	6	8	9
Massachusetts	12	17	13	0	6	4	5
Michigan	12	18	6	6	0	10	6
New Jersey	9	14	8	4	10	0	4
Pennsylvania	8	13	9	5	6	4	0

There are also fixed costs of contracting (paper, personnel, and so forth) incurred by each arsenal (total cost), as follows:

	Total contracting cost
Raritan Arsenal	$30
Redstone Arsenal	$52
Rock Island Arsenal	$60

Find the proper manner of letting contracts so as to cost the government as little as possible. What is this cost? (This problem is continued in Problem 6-6.)

3-7 Exercise in alternate optima. Given the following matrix after step 11,

	9	6	4	4
6	6	4	0	0
5	5	0	0	0
12	0	0	0	3

(*a*) Find all basic optimal solutions.

(*b*) Find all nonbasic optimal solutions with integer values of the x's.

Some Extra Drill Problems

Other transportation problems will be found at the end of Chapter 6; ignore the text and use the matrices. You can also make up your own problems (be sure only that $\Sigma\alpha_i = \Sigma\beta_j$); see comment on extra drill problems in Chapter 2.

Advanced Problems

3-8 You must store m substances (say, different grains, like wheat and barley, or different chemicals which you have manufactured) in n different storage facilities. The quantities to be stored and the storage capacities are known. Certain substances require controlled environments (temperature, humidity) which are more or less easily provided at different facilities; for this reason, and because of varying densities, and so forth, there is a specific storage cost for one unit of the ith substance in the jth facility for one time period. These costs are also known.

(*a*) Formulate as a transportation problem.

(*b*) Why do you suppose this model of this problem might not be very useful in practice? This problem is continued in Problem 6-14, but please don't peek there until you have answered it.

3-9 (*a*) In step 8, do you always subtract more than you add?

(*b*) The last time you went through step 8 in Problem 3-1, you added one to columns 1 and 4 and subtracted one from row 3. Did you still subtract more than you added?

(*c*) Interpret your answer to part (*a*) in terms of model (3-1) or (3-3).

4

THE GENERAL LINEAR PROGRAMMING PROBLEM

We have already learned a good deal about the assignment and transportation problems: Each is a problem of choosing values of certain variables which minimize a cost subject to certain constraints; that is, each is an example of optimization under constraints. At the end of the last chapter we formulated these problems (and also their duals) in this way and indicated that certain generalizations of these problems would be called linear programming. We therefore now define the generalized problem as follows: A *linear programming problem* consists of finding values of a set of variables which maximize or minimize a linear function of those variables subject to constraints in the form of linear equations or linear inequalities in those variables.

Note first that such problems are decision problems; we are asked to choose values of the variables. However, we are not completely free in our choices; there are certain constraints, in the form of either equations (for example, the number of units shipped from the first origin must exactly equal the number of units available at the first origin) or inequalities (for example, we must not subtract more from an element in the cost matrix than there is initially, lest we reduce it to a negative number). Furthermore, these constraints must be linear; complicated constraints involving such things as the squares of vari-

	Steak (pound)	Carrots (bunch)	Potatoes (bushel)	Vodka (fifth)	Requirement
Protein	4	$1\frac{1}{2}$	$\frac{1}{2}$	$\frac{1}{4}$	17 (oz)
Carbohydrate	1	2	87	5	49 (oz)
Fat	3	0	0	0	6 (oz)
Vitamin A	1000	30,000	2000	10,000	20,000 (units)
Vitamin B1	33	2	0	147	82 (mg)
Cost	189	57	100	349	(cents)

FIGURE 4-1
The diet problem.

ables or the products of variables are not permitted. Finally, we are asked to maximize a profit or minimize a cost, which is also expressed as a linear function of these variables. "Linear function" means that if one unit of a particular variable is worth \$3, then every unit of that variable is also worth \$3. A nonlinear function might be one in which there were, for example, economies of scale so that the more of something we produced the more efficiently we would produce it. Or there might be *synergy*, a special kind of interaction such that the total effect of several products or activities is different from the sum of the effects of the individuals. Such functions are nonlinear and are more difficult to deal with. This book deals with the simpler linear programming problems; nonetheless, we shall develop many techniques which are quite useful in the real world.

We shall first indicate how such problems arise by giving two examples; there will be numerous further examples as the book proceeds. Then we will develop general methods for solving these problems. It turns out, however, that that's not as hard as you think; you already know almost all you need to know to solve the general linear programming problem.

4.1 APPLICATIONS OF LINEAR PROGRAMMING

We shall now develop two examples of how linear programming problems arise. In one, we will be asked to minimize a cost; it is called *the diet problem* and will be referred to frequently as a typical minimization problem. In the other, we are asked to maximize a profit; it is called *the product-mix problem* and will be referred to frequently as a typical maximization problem.

	x_1	x_2	x_3	x_4		
	4	$1\frac{1}{2}$	$\frac{1}{2}$	$\frac{1}{4}$	$\geq$	17
	1	2	87	5	$\geq$	49
	3	0	0	0	$\geq$	6
	1000	30,000	2000	10,000	$\geq$	20,000
	33	2	0	147	$\geq$	82
Minimize	189	57	100	349	$=$	z

FIGURE 4-2
Mathematical model of the diet problem.

4.1.1 The Diet Problem

The problem is illustrated in Figure 4-1. We have a number of foodstuffs, such as steak, carrots, potatoes, and so forth, and a number of dietary constituents, such as protein, carbohydrate, fat, vitamin A, vitamin B1, vitamin B2, and so forth. While we do not vouch for the numerical values of any of the examples given here, we assume that we are given all the necessary data concerning the amounts of these dietary constituents in these foodstuffs. For example, each pound of steak contains, according to Figure 4-1, 4 ounces of protein, 1 ounce of carbohydrate, 3 ounces of fat, 1000 units of vitamin A, 33 milligrams of vitamin B1, and so forth. We are also given the cost—for example, \$1.89 per pound for steak—and the requirements over some stated period of time—for example, at least 17 ounces of protein per week. Note that the units need not be the same for one foodstuff and another or for one requirement and another, although within any one column or any one row the units must be consistent.

Now our problem is the following: We wish to select a diet which gives us at least the minimum weekly requirements of each of the dietary constituents and which still costs as little as possible.

To build a mathematical model of this problem, which will turn out to be a linear programming problem, we must first define a set of variables. We ask you to pay special attention to this step, since it has been our experience that it is one that causes students the most difficulty. You have probably had this difficulty with "word problems" all the way along, starting way back when 3 + 5 wasn't too hard, but if Mary had 3 apples and Jane had 5 apples it got tougher. In the formulation of a linear programming problem—in fact in the formulation of almost any mathematical model—the first step, and perhaps the most important step, is selection and careful definition of the variables. In this case the selection is indicated in Figure 4-2. We let x_1 equal the number of pounds of steak that we shall purchase per week; let x_2 equal the number of bunches of carrots that we shall purchase per week; and so forth.

Now if we do purchase exactly x_1 pounds of steak, how many ounces of protein will we get? According to the given data, each pound of steak supplies

4 ounces of protein, and therefore x_1 pounds of steak will supply $4x_1$ ounces of protein. Similarly, x_2 bunches of carrots will supply $1.5x_2$ ounces of protein. And when we have bought all the indicated amounts of all the foodstuffs, the total amount of protein we will have will be $4x_1 + 1.5x_2 + 0.5x_3 + 0.25x_4 + \cdots$. This amount of protein must be at least equal to, if not greater than, 17 ounces. This leads to an inequality constraint of the following form:

$$4x_1 + 1.5x_2 + 0.5x_3 + 0.25x_4 + \cdots \geq 17$$

Similarly, as indicated in Figure 4-2, each row of this matrix leads to an inequality constraint of the "greater-than-or-equal-to" type. Finally, the total cost of our diet (in cents) will be

$$189x_1 + 57x_2 + 100x_3 + 349x_4 + \cdots = z$$

and this quantity is to be minimized. If you will check this problem against our definition of a linear programming problem above, you will see that it fits precisely. Note, in particular, the linearity both in the constraints and in the objective function (remember that the objective function is the name of the quantity to be minimized—in this case z). In particular, every pound of steak supplies exactly 4 additonal ounces of protein regardless of how many other pounds of steak we have or how much of anything else we have. Similarly, each additional pound of steak costs exactly \$1.89 regardless of anything else, and similarly for all the other foods.

If this were a real problem, in which there might be 37 dietary constituents and 142 different foodstuffs, then it would be possible to solve it by the methods we are shortly going to learn; although it would be too difficult to do by hand, a large digital computer could do it in a few seconds. We need hardly warn you that if you do choose to solve this problem numerically, it would be advisable to get better data than we have supplied, some of which is clearly based on wishful thinking.

4.1.2 The Product-Mix Problem

You, the president of the Widget Manufacturing Company, have to decide exactly what to manufacture next month. You make several products, among which are widgets, gidgets, yidgets, and zidgets, as indicated in Figure 4-3. At the bottom is indicated the unit contribution of each of these products; for example, for each 1000 widgets manufactured you make \$189 profit. For purposes of this problem we assume no marketing constraints; that is, you are capable of selling everything you make at the stated profits. However, you have certain limitations due to limited availability of raw materials and limited production facilities. For example, there are only 17 tons of steel in the factory. (We remind you once again that these numbers are not to be taken seriously; if it would make you happier, you can assume that the 17 represents 17 pounds or 17,000 tons of steel; however, we shall continue to refer to it as

	Widgets (1000's)	Gidgets (100's)	Yidgets (1000's)	Zidgets (gross)	Available
Steel	4	$1\frac{1}{2}$	$\frac{1}{2}$	$\frac{1}{4}$	17 (tons)
Copper	1	2	87	5	49 (tons)
Lumber	3	0	0	0	6 (board feet)
Labor	1000	2000	20,000	10,000	20,000 (man-weeks)
Milling machine	33	2	0	147	82 (hours)
Profit	189	57	100	349	(dollars)

FIGURE 4-3
The product-mix problem.

17 tons for the purposes of this illustration.) Now each 1000 widgets that are produced use up 4 tons of this precious steel; or, if you prefer, each widget consumes 0.004 ton of steel, and similarly for all the other numbers in the matrix. For example, it takes 10,000 man-weeks of labor to make a gross of zidgets, and we have only 20,000 man-weeks of labor available.

The mathematical formulation of this *product-mix problem* is, of course, similar to the last one. First we define a set of variables. This time we let x_1 equal the number of thousands of widgets to be produced; x_2 will be the number of hundreds of gidgets; and so forth. The total amount of steel consumed will then be

$$4x_1 + 1.5x_2 + 0.5x_3 + 0.25x_4 + \cdots$$

and this amount of steel must at most equal 17, although it may be less. Thus, our first constraint becomes the inequality

$$4x_1 + 1.5x_2 + 0.5x_3 + 0.25x_4 + \cdots \leq 17$$

and similarly for each other constraint, as shown in Figure 4-4. The bottom line now represents total profit, z, and our task will be to maximize this, subject to the stated constraints.

We might, of course, have called the variables of the second problem by some other name such as y's, but you might as well get used to the fact that the variables are almost always called x's. We will introduce y's at a later time, but they will refer to the variables of the dual problem. Each of these problems does, in fact, have a dual (obtained by writing each row as a column or, if you prefer to look at it that way, each column as a row), but we will delay consider-

	x_1	x_2	x_3	x_4		
	4	$1\frac{1}{2}$	$\frac{1}{2}$	$\frac{1}{4}$	$\leq$	17
	1	2	87	5	$\leq$	49
	3	0	0	0	$\leq$	6
	1000	30,000	2000	10,000	$\leq$	20,000
	33	2	0	147	$\leq$	82
Maximize	189	57	100	349	=	z

FIGURE 4-4
Mathematical model of the product-mix problem.

ation of these dual problems until we have learned more about the primal problems, including how to solve them.

We also remind you (see pages 19 to 20) that, if there are 142 variables in this problem with 37 constraints, then we can consider it to be represented geometrically by 37 hyperplanes in 142-dimensional space. We have already admitted our inability to graph such situations. We shall turn now to problems where there are only two variables so that it will be easy to graph them and exceedingly easy to solve them. While we're at it, however, we will gain a good deal of understanding about the nature of such problems, and we will develop an algorithm which we can then use for solving large problems.

4.2 STARTING THE SOLUTION PROCESS

We shall work with the following small problem and variations on it:

Maximize: $$z = 2x_1 + 3x_2$$

Subject to:
$$\begin{aligned} 2x_1 + x_2 &\leq 5 \\ x_1 + 2x_2 &\leq 5 \end{aligned} \qquad (4\text{-}1)$$

This problem is in the same form as Figure 4-4, the product-mix problem. If you would like to think of it as a specific application while we are working with it, then we shall let x_1 equal the number of thousands of widgets to be produced tomorrow, and x_2 will equal the number of hundreds of gidgets to be produced tomorrow. Note the care with which we have defined these variables; it is not adequate to say x_1 = widgets and x_2 = gidgets. We are left in doubt as to the units in which they are being measured, whether we are talking about the number of widgets to be produced or the number to be sold, and whether the period under consideration is a day or a week. You may also consider that the first constraint concerns the usage of steel and the amount that is available (5 tons) and that the second constraint might concern labor, with 5 man-days available. Finally, z is measured in dollars. In actually solving the problem, the meanings of all these numbers are not significant; but when it comes to interpreting the solution they will be important, and so we must be specific about them when we formulate the mathematical model.

In looking for ways to solve this problem, that is, to find the values of the x's which maximize the objective function subject to the constraints, we note that we don't really know very much about inequalities, but we know a good deal about equations. (If you don't know a good deal about equations, you had better restudy Chapter 1.) Before proceeding any farther, therefore, we shall change the problem so that it is in the form of equations. While we are at it, we shall make a couple of other changes which convert the problem into what is called standard form. We now define this form. A *linear programming problem in standard form* consists of finding nonnegative values of a set of variables which minimize a linear function of those variables subject to constraints in the form of linear equations in those variables. There are three differences between the standard form and the general linear programming problem: (1) In the general problem the variables might be allowed to take on any values; in the standard form the values of the variables must be nonnegative (that is, either 0 or positive). (2) In the general linear programming problem we might be asked to either maximize or minimize the objective function; in the standard form we are asked always to minimize. (3) In the general linear programming problem the constraints may be in the form of equations or inequalities; in the standard form only equations are permitted. We now discuss each of these three restrictions in turn.

In many problems the restrictions to nonnegativity are inherent. For example, it is "obvious" that we cannot manufacture a negative number of widgets. The things which are "obvious" are not always valid. Suppose widgets used an enormous amount of resources, say, 100 tons of steel and 100 man-weeks of labor, and yielded a very small profit, say, \$1; then it is obvious that we should not make any widgets in our optimum product mix. But you have to tell the computer that; if you don't tell the computer that, it will try various negative values of x_1, which is equivalent to buying widgets on the market at \$1 per unit and then breaking them down to yield 100 tons of steel and 100 available man-hours of labor, which can then be used to make other more profitable products. Obviously this makes no sense in the real world, but it makes perfectly good mathematical sense if we do not specify explicitly in the mathematical model that it is to be prohibited. Thus, we must add as additional constraints to the problem (4-1)

$$x_1 \geq 0 \qquad x_2 \geq 0$$

It is not always true that real problems are restricted to nonnegative variables. In fact, in our limited experience we have already seen one linear programming problem in which the optimal value of at least one of the variables was negative. Specifically, in the dual of our example transportation problem, the optimal value of u_2 turned out to be -2. This negative value of this particular variable has important practical significance, in terms of things called opportunity costs and shadow prices, about which we shall have more to say later. If we do have a real problem in which the decision variables might take on negative values, then we must use some trick to convert the problem into standard form. There are several different ways of doing this, but since our present problem does not require them, we shall defer consideration of them.

As we saw in Section 2.6, conversion of a maximization problem to a minimization problem, and vice versa, is a trivial matter. We need merely to multiply the objective function by -1; for purposes of standardization, therefore, it doesn't really matter very much which we choose. W shall choose always to minimize because this leads to certain desirable symmetries, as explained at the end of Section 5.2. In particular, for this problem we change the objective function from

Maximize: $z = 2x_1 + 3x_2$

to

Minimize: $-z = -2x_1 - 3x_2$

Thus we convert the objective function (something to be maximized or minimized) into a *cost function* (something to be minimized).

We convert from inequalities to equations by adding something called a slack variable. Consider the first of the inequalities above:

$$2x_1 + x_2 \leq 5$$

We now assert that the following equation is exactly equivalent:

$$2x_1 + x_2 + s_1 = 5$$

s_1 is called a *slack variable*, this being the name given to a variable introduced for the purpose of converting an inequality to an equation. Why is it equivalent? Because any nonnegative values of x_1 and x_2 which satisfy the inequality also satisfy the equation, and vice versa, provided only that we insist that s_1 also be nonnegative. Try it. Let $x_1 = 1$ and $x_2 = 1$; do these values satisfy the inequality? Yes. What happens when we plug them into the equation? They yield $s_1 = 2$. Try another set of values, $x_1 = 2$ and $x_2 = 2$. Do these values of the x's satisfy the inequality? No, because $2x_1 + x_2$ is now equal to 6 and this is greater than 5. What happens when we plug them into the equation? They yield $s_1 = -1$, which does not satisfy the nonnegativity requirement. In fact, as you can see, any pair of values of x_1 and x_2 which satisfy the inequality give a nonnegative s_1 when they are plugged into the equation. If they fail to satisfy the inequality, then they give a negative s_1 when they are plugged into the equation. Looking at it the other way, any set of nonnegative values of x_1, x_2, and s_1 which satisfy the equation also satisfy the inequality.

The word "slack" is not new. If $x_1 = 1$ and $x_2 = 1$, then two units of the resource are left over, or constitute the slack on the resource. In fact, the slack variable s_1 measures precisely how much slack we have on this first resource.

If s_1 is equal to 0, then we have no slack; the resource has been exactly used up and the inequality is exactly satisfied (as an equation). If s_1 should come out negative, it would mean there was negative slack—we have used up more of the resource than we have available, and we have violated the constraint.

Similarly, the second inequality $x_1 + 2x_2 \leq 5$ becomes the second equation $x_1 + 2x_2 + s_2 = 5$. Note that a different slack variable is required for each constraint; since the number of tons of steel left over may be different from the number of man-days of labor left over, these numbers must be represented by different slack variables. In general, we shall have to add one slack variable for every inequality constraint in a problem, and when we have done this there will no longer be any inequalities in the problem, and all constraints will be in the form of equations.

Although it will often be convenient to refer to the *i*th slack variable as s_i as we have above, it will also often be convenient to designate the slack variables by x's. In fact, the whole idea of putting them in is to get ordinary equations similar to those which we studied in Chapter 1. Therefore, we will now introduce two variables $x_3 = s_1$ and $x_4 = s_2$ so that our problem becomes

$$\begin{aligned} 2x_1 + x_2 + x_3 \qquad &= 5 \\ x_1 + 2x_2 \qquad + x_4 &= 5 \\ \text{Minimize:} \qquad -2x_1 - 3x_2 \qquad &= -z \end{aligned}$$

which, if we detach coefficients, becomes

$$\begin{array}{rrrrr} 2 & 1 & 1 & 0 & 5 \\ 1 & 2 & 0 & 1 & 5 \\ -2 & -3 & 0 & 0 & -z \end{array} \qquad (4\text{-}2)$$

It is assumed, although we do not write it, that $x_j \geq 0$, $j = 1, \ldots, 4$ (which means $x_1 \geq 0$, $x_2 \geq 0$, $x_3 \geq 0$, and $x_4 \geq 0$). Since we have called x_3 and x_4 "slack variables," we now need a special name for x_1 and x_2, and we shall call them *structural variables*; this name will always refer to the real variables of the problem—the ones that we started with when we built the mathematical model. Note that the slack variables are given zero-cost coefficients. It may be that if steel is left over we can sell it, so that c_3 (the coefficient of s_1 in the cost function) should be negative, or that workers who are not kept busy will get into mischief, so that c_4 (the coefficient of s_2 in the cost function) should be positive. But these are not the assumptions of this problem—we have assumed that we are indifferent between using 5 tons of steel and using less than five; we are forbidden only to use more than five.

Having put the problem in standard form, we adopt a standard notation shown in Figure 4-5. Thus, the first two constraint equations would be

$$\begin{aligned} a_{11}x_1 + a_{12}x_2 + \cdots &= b_1 \\ a_{21}x_1 + a_{22}x_2 + \cdots &= b_2 \end{aligned}$$

Here, for example, $a_{11} = 2$, $a_{12} = 1$, $a_{13} = 1$, $a_{14} = 0$, $b_1 = 5$, and so forth. This

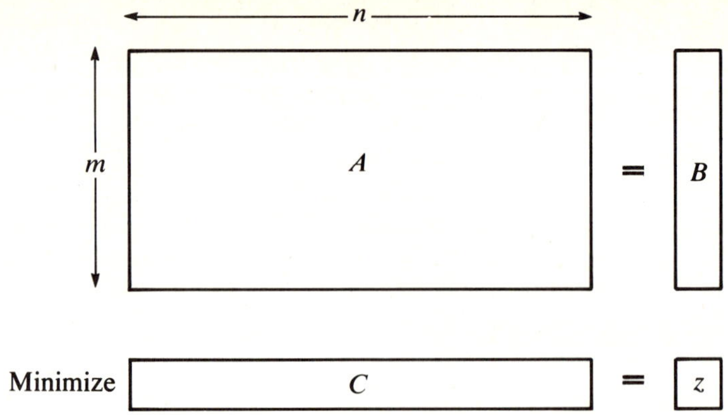

FIGURE 4-5
A linear programming problem in standard form.

can be written

$$\sum_j a_{1j}x_j = b_1$$

$$\sum_j a_{2j}x_j = b_2$$

If there are m such constraint equations, we may write

$$\sum_j a_{ij}x_j = b_i \qquad i = 1, \ldots, m$$

or

$$AX = B$$

as discussed on page 15. The objective function, or cost function, is now

Minimize: $$c_1x_1 + c_2x_2 + \cdots = z$$
or
Minimize: $$CX = z$$

And we have $x_j \geq 0, j = 1, \ldots, n$, or, in vector form, $X \geq 0$. Finally, we may summarize any linear programming problem in standard form by this extremely compact notation:

Minimize: $$CX = z$$

Subject to: $$AX = B$$
and $$X \geq 0$$

where A is an $m \times n$ matrix, B, C, and X are vectors, and z is a number. We shall call B the *requirements vector*, C the *cost vector*, and A the *activity ma-*

trix; thus each b_1 is a *requirement*, each c_j is a *cost coefficient*, each a_{ij} is an *activity coefficient*, and z is the *cost*. To be rigorous, we should have introduced a different symbol here, because this z is the same as the previous $-z$ (and it would be invalid to write $z=-z$); but we are trying to use as few symbols as possible.

Now if you examine the problem as stated in (4-2), you will note that it is not only in standard form but also in the "canonical form" we worked so hard to get by Gaussian elimination back in Chapter 1. The first two lines of this problem (the A and B parts) represent two constraint equations in four variables, and in these two equations we have our permutation matrix in columns 3 and 4. That is, these two columns constitute a square matrix with exactly one 1 in each row and one 1 in each column and everything else 0. This means that if we look at a particular basic solution (that is, a solution in which two of the variables are equal to 0), namely, the solution in which $x_1 = x_2 = 0$, we see a solution to this set of equations immediately displayed: $x_3 = 5$, $x_4 = 5$, and $-z=0$. This is clearly not the optimal solution to the problem. However, it is a *feasible solution* to the problem; that is, it is a solution that satisfies all the constraints. This solution is: Produce no widgets and no gidgets, have 5 tons of steel and 5 man-days of labor left over, and make no money. Believe it or not, this is a very useful starting point for getting to the optimal solution of our problem. We call your attention to the two uses we have made of the word solution: A *feasible solution* is any set of nonnegative values of the x's which satisfy the constraints. In general, since $n > m$ (that is, since the number of variables exceeds the number of equations), there will be an infinite number of such feasible solutions. An *optimal feasible solution* is a feasible solution which minimizes the cost function.

Now what makes it comparatively easy to go from the feasible solution to the optimal feasible solution is a remarkable theorem discovered by the American mathematician George Dantzig, which says roughly that the optimal solution is always a basic solution (page 32). We shall show shortly that this theorem is not only valid, but obvious. For the moment we ask you to note that this theorem indicates how our algorithm for solving linear programming problems will work:

Step 1 Put the problem in standard form. Go to step 2.

Step 2 Put the problem in canonical form, that is, a form like (4-2) containing appropriate columns of 1s and 0s. Go to step 3.

Step 3 Examine the basic solution obtained by letting all the nonbasic variables go to zero. If this solution is optimal, stop; otherwise go to step 4.

Step 4 Make one nonbasic variable basic by Gaussian elimination, in such a way as to reduce the cost function. Go to step 3.

While the above is not precise, you already know how to do most of it. For example, you know how to convert the problem into standard form by adding slack variables and by converting from a maximization to a minimization if necessary. There are some other details about nonnegativity which we will cover later. Step 2 is automatic in problems like the one we are now examining, since we note that (4-2) is already in canonical form. If it should not be in this form, it can be put into this form by Gaussian elimination. Step 3 we will learn about below, but it is really rather easy. As for step 4, once we have decided which new variable should become basic, we have already learned in Section 1.4 how to develop the new set of equations. However, before we proceed any further with this problem, we pause to draw a graph which represents it, and to indicate why Dantzig's Theorem is true.

4.3 THE GEOMETRIC ANALOGY

In graphing the problem (4-1) we note that we have never before graphed an inequality. To do this we first pretend that it is an equation; thus in Figure 4-6 we first draw a graph of the line $2x_1 + x_2 = 5$. Since we are interested only in nonnegative x's, we draw only that portion of the line lying between the axes. As we have pointed out in Section 0.4, every point on this line corresponds to a pair of x's which satisfies this equation, and vice versa. Every point above and to the right of this line represents a pair of x's such that $2x_1 + x_2 > 5$, and every point below and to the left of the line represents a pair of x's for which $2x_1 + x_2 < 5$. Or, to say it another way, points above and to the right of the line violate our first constraint; points below and to the left of the line satisfy the first constraint. Thus, we have shaded in Figure 4-6 a region which includes every possible pair of nonnegative values for x_1 and x_2 which satisfy our first constraint.

Note again the equivalence of the slack variable form. Points inside the shaded region of Figure 4-6 correspond to positive values of the slack variable x_3 (or s_1). Points on the diagonal line bounding this region represent $x_3 = 0$. Points on the vertical and horizontal lines bounding the region represent places where $x_1 = 0$ and $x_2 = 0$, respectively.

Now in Figure 4-7 we add the line corresponding to our second constraint and shade a region representing x-values which satisfy this constraint. A point (that is, a pair of values of x_1 and x_2) satisfies all the constraints of this problem if and only if it falls in the crosshatched (doubly shaded) region. This crosshatched region is called the *region of feasible solutions*. Anything inside this region is *feasible*; that is, it satisfies all the constraints of the problem. Anything outside the region is infeasible; that is, it violates at least one of the constraints of the problem. At this point in our development, we have four constraints, each represented by one of the lines that bounds the crosshatched region. A violation of any one of these constraints implies that one of our four variables will be negative. The points inside the crosshatched region have all four of the variables positive. On any one of the four lines that bounds the region, one of these four variables equals 0; that is, on the horizontal line (or x_1-axis) we have $x_2 = 0$, on the vertical line (or x_2-axis) we have $x_1 = 0$, on the first line that we drew in Figure 4-6 we have $x_3 = 0$, and on the last line that we drew in Figure 4-7 we have $x_4 = 0$. Finally, this four-sided polygon which bounds the region of feasible solutions has four *vertices*, or corners. Each *vertex* is a point where two of these lines meet and therefore a point where two of the variables are equal to zero.

Now finally we wish to graph the objective function $z = 2x_1 + 3x_2$. If z were some particular number like 6, we would know how to graph it; so we let $z = 6$, and we draw the corresponding line on Figure 4-8. Every point on this line represents a mixture of products which leads to a profit of 6 units of money ($6). For example, we might make no widgets ($x_1 = 0$) and 200 gidgets ($x_2 = 2$) which is a feasible program—that is, a feasible product mix. The point $x_1 = 3$, $x_2 = 0$ also lies on this line and would yield a profit of $6 if we could make it, but it represents an infeasible product mix. It is clear that there are several ways in which we could make a profit of $6, but perhaps we could make more; perhaps we could make $12. Let us try the graph of $z = 12$; that is, $2x_1 + 3x_2 = 12$. This line is also on Figure 4-8, and we note that no point on this line is feasible; that is, there is no intersection between this line and the region of feasible solutions.

Thus, in order to graph the objective function, we first have to pick a particular value of z; each such value of z leads to a line, and all such lines have the same slope, that is, they are all parallel. What we should like to do is to take a line such as the one for $z = 6$ and push it parallel to itself, up and to the right as far as possible; but we cannot push it as far as the line $z = 12$ because there is no feasible point on this line. The farthest we can push it is $z = \frac{25}{3}$; as you will note from Figure 4-8, this line has exactly one point in common with the region of feasible solutions, namely, the point $x_1 = \frac{5}{3}$, $x_2 = \frac{5}{3}$. Any point of higher profit is infeasible, and no other feasible point gives a profit as high as this; thus, this point is "the solution" to our problem—that is, the optimal feasible solution.

Now we have found graphically a solution to this particular problem; we have also indicated something fundamental about the solution to any linear programming problem; namely, that the region of feasible solutions will always be a polygon and that the optimal solution will always be at a vertex. We shall have to extend our definition of "polygon" a little, however. For example, an unbounded region, which goes on to infinity in some directions, may be a polygon; certain three-dimensional objects, such as a cube or pyramid, are more properly called *polyhedra*, but we shall call them polygons; and a "hyper-polyhedron" in higher dimensional space is also, for our purposes, a polygon.

In this particular problem the solution was at one particular vertex, but now suppose that the objective function had been different. If we had received a profit of $1 each for gidgets, and 0 for widgets, then our z-lines would have been a set of horizontal lines; and when we moved these up as far as possible,

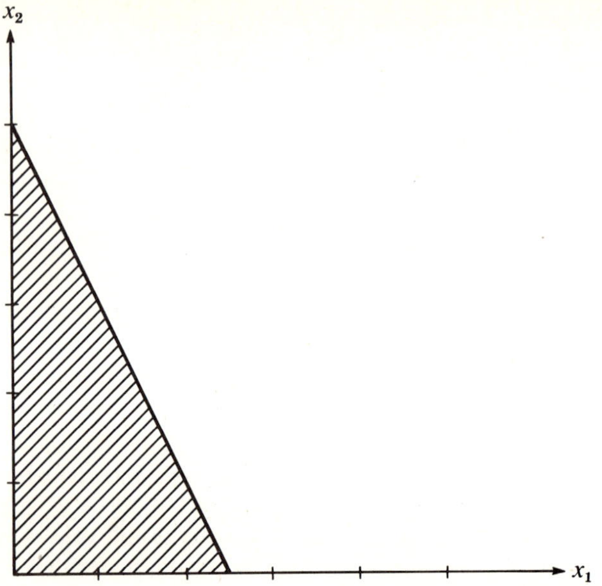

FIGURE 4-6
Region of feasible solutions for $2x_1 + x_2 \leq 5$, $x_1 \geq 0$, $x_2 \geq 0$.

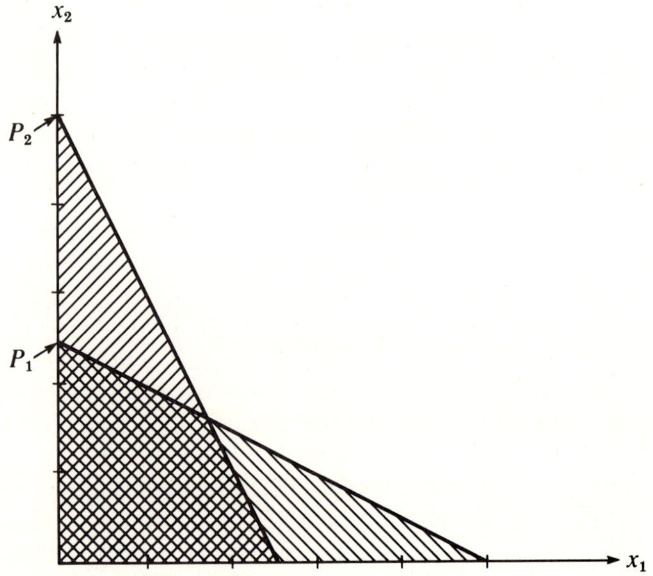

FIGURE 4-7
Region of feasible solutions for (4-1).

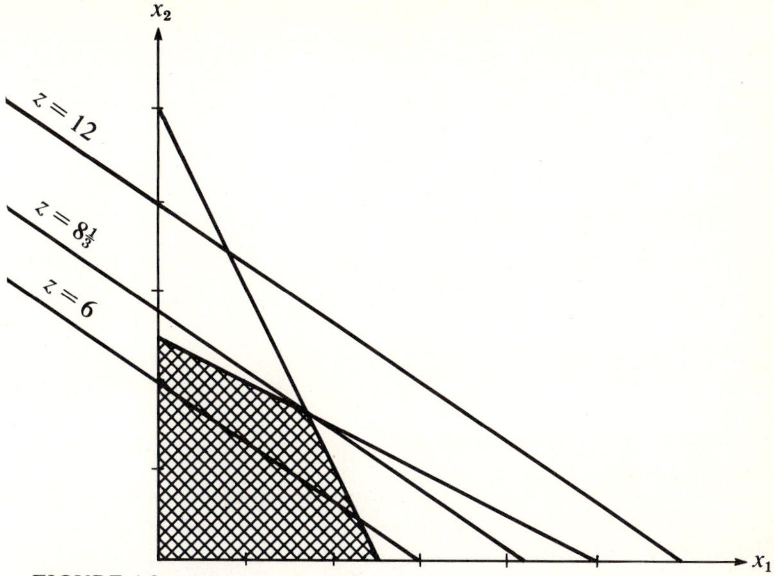

FIGURE 4-8
Graph of the problem (4-1).

our solution point would have been $x_1 = 0$, $x_2 = \frac{5}{2}$. A different vertex, indeed, but still a vertex. Suppose there had been an additional constraint, say, copper in addition to steel and labor; then there would have been an additional line on the diagram, and our region of feasible solutions might have been bounded by a five-sided polygon instead of a four-sided one. Nonetheless, when we moved our objective-function line as far as possible, we would eventually find our solution point at a vertex.

We have just made a geometric assertion about the nature of the optimal feasible solution to a linear programming problem; now we wish to make the corresponding algebraic assertion. We have explained that each vertex is a point at which two of the four variables are equal to zero, because each line represents points where one of the variables (either a structural variable or a slack variable) is equal to zero, and a vertex is the intersection of two lines. Thus, equivalent to the geometric statement that the optimal feasible solution must lie on a vertex is the algebraic statement that in the optimal feasible solution two of the variables must be equal to zero. But this is precisely what we have called a basic solution. We have formally defined a basic solution, when there are m equations in n variables, as being one in which $n - m$ of the variables take on the value 0. Consider all problems which have two structural variables (say, widgets and gidgets or meat and potatoes); these are the problems which can be represented by an ordinary two-dimensional graph. If we have m inequality constraints, we will have to add one slack variable to each, so our total

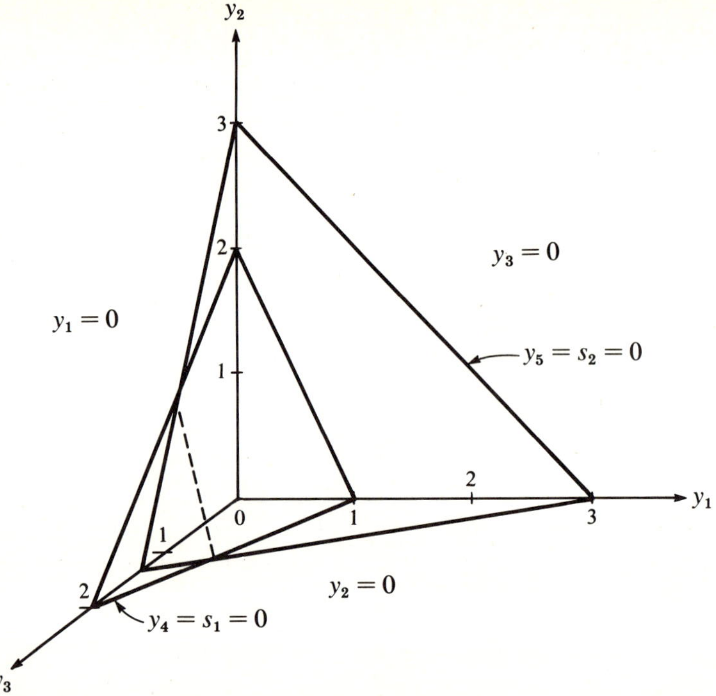

FIGURE 4-9
A three-dimensional linear programming problem.

number of variables will be $n = m + 2$ (m slack variables, 2 structural variables); thus $n - m = 2$. Therefore each vertex (where 2 variables are zero) corresponds to a basic solution (where $n - m$ variables are zero).

To illustrate this point once more, we now graph a three-dimensional problem. This is a minimization problem (that is, one like the diet problem) and consists of two constraints (dietary requirements) in three variables (foodstuffs) as follows:

$$2y_1 + y_2 + y_3 \geq 2$$
$$y_1 + y_2 + 2y_3 \geq 3$$

Minimize:
$$5y_1 + 3y_2 + 5y_3 = w$$

We have graphed this problem in Figure 4-9; each constraint inequality becomes a plane, and there are two of them. Because the constraints are of the greater-than-or-equal-to variety, the region of feasible solutions lies above the constraint planes and is unbounded (we continue to call this unbounded three-dimensional region a polygon). The cost function consists of a set of parallel planes, and we would like to move these down as far as possible toward the origin. You can see that the exact point where the solution occurs depends upon the slope or inclination of the plane representing the objective function. However, it must eventually come to rest on one of the corners or vertices. Each such vertex is a place where three planes intersect. Each plane represents a point where one of the five variables is zero (we started with three structural variables, and we added two slack variables to convert the two constraint inequalities into equations). On Figure 4-9 we have indicated which variable is equal to zero for each plane; for example, the plane of the paper, that is, the plane in which the y_1- and y_2-axes are both lying, is the plane in which $y_3 = 0$. Note that in this problem, after we have reduced it to standard form, we have $m = 2$ equations and $n = 5$ variables. It follows that $n - m = 3$, and a basic solution will have three variables equal to zero. We have shown that the optimal feasible solution must be one in which three of the variables are zero; so once again we have shown that the optimal feasible solution must be a basic solution.

We can now state *Dantzig's Theorem* formally: If a linear programming problem in standard form has a unique optimal feasible solution, it will be a basic solution, or, equivalently, it will be at a vertex. And if it has alternate-optimal feasible solutions, at least one will be basic. We are sorry about the irritating "if's," but it is necessary because some linear programming problems don't have optimal feasible solutions. This can arise in either of two ways. Sometimes there are no feasible solutions (the constraints are too constraining), and sometimes the solution set is unbounded (no matter how good a solution you find, there is always a better one). We will worry about insoluble problems later (Section 4.10). Right now we have to find out how to solve soluble problems.

4.4 DEVELOPING THE ALGORITHM

We repeat here our example problem:

$$\begin{array}{rrrrr} 2 & 1 & 1 & 0 & 5 \\ 1 & 2 & 0 & 1 & 5 \\ -2 & -3 & 0 & 0 & -z \end{array} \tag{4-2}$$

If we let all the nonbasic variables become zero (the basic solution), our equations read

$$\begin{aligned} 0 + 0 + x_3 + 0 &= 5 \\ 0 + 0 + 0 + x_4 &= 5 \\ 0 + 0 + 0 + 0 &= -z \end{aligned} \tag{4-3}$$

or

$$x_1 = 0,\ x_2 = 0,\ x_3 = 5,\ x_4 = 5,\ z = 0$$

We know this is a feasible solution, and we suspect that it is not an optimal solution. We know that if an optimal feasible solution is to exist, it must be a basic solution. The question is: Is there some basic solution which can be easily reached from this one and which is obviously better? And the answer is yes. If we are to get a new basic solution, then one of the nonbasic variables (which at present are x_1 and x_2) must become basic, that is, must increase from 0 to some positive value, and one of the basic variables (which at present are x_3 and x_4) must become nonbasic, that is, must decrease from a positive value to zero.

In case you are confused by now about basic and nonbasic, we repeat: A *basic variable* has one coefficient of 1 and all other coefficients (in all other rows) of 0. A *basic solution* has at least $n - m$ variables equal to zero (and if it is degenerate, more than $n - m$). The basic solution in which we are interested is the one in which each nonbasic variable (there are always exactly $n - m$ of them) equals zero. Note that one kind of variable (basic) has coefficients mostly zero, and the other kind of variable (nonbasic) has value zero. This may be confusing, but it makes the arithmetic delightful, because almost every product is then zero. Remember that in the transportation problem either the cost or the quota was zero, so every term in the sum $\Sigma\Sigma\, c_{ij}x_{ij}$ was zero. Here almost every term in the sum $\Sigma\, a_{ij}x_j$ is zero; and in the one remaining term $a_{ij} = 1$, so that each equation reduces to $x_j = b_i$. The whole set of equations then looks like (4-3), which is easy.

Now suppose we let x_1 become basic, what would happen to the cost function? According to (4-2) the cost, $-z$, would decrease by two dollars for each unit by which x_1 increased; that is the meaning of the coefficient $c_1 = -2$ in the cost function. Such a cost decrease is good. But in addition, x_3 and x_4 are going to change. What effect will this have on the total cost? Answer: None. The coefficients of each of these variables is zero in the cost function, and changing them will not change the total cost. A similar argument indicates that if x_2 were to become positive, we would improve (that is, reduce) the cost function by \$3 for each unit by which x_2 was increased. Now which one should we work with? That amounts to asking whether you would rather have an improvement by \$3 per unit or an improvement by \$2 per unit. Other things being equal, we would choose 3, and so we will choose to let x_2 be the incoming variable, that is, make x_2 basic.

Let us understand what this means. At the present time x_1 and x_2 are nonbasic variables and therefore are equal to zero. x_3 and x_4 are basic variables and happen each to be equal to 5 at the moment. We are going to allow x_2 to increase; we are not yet sure how much it will increase, but we would like to increase it as much as possible, because we are going to get a net improvement of \$3 for each unit of x_2 that we bring in. In the new basic solution, then, x_2 will be basic, and x_1 will remain nonbasic. Of the other two variables, x_3 and x_4, one of them will remain basic, and one of them will become nonbasic. We will find out which one when we find out how much x_2 may increase.

The first equation of (4-2) reads

$$2x_1 + x_2 + x_3 = 5$$

2	1	1	0	5	5
1	2	0	1	5	$\frac{5}{2}$
-2	-3	0	0	$-z$	

FIGURE 4-10
How to find the variable which will become nonbasic.

But since we know that x_1 is zero and is going to remain zero, we can omit it:

$$x_2 + x_3 = 5$$

Solving this equation for x_3 is easy because of the canonical form:

$$x_3 = 5 - x_2 \qquad (4\text{-}4)$$

Now we wanted to make x_2 as big as possible; what does this equation tell us about how big we can make it? It says clearly that if x_2 remains smaller than 5, x_3 will remain positive; but if x_2 becomes larger than 5, x_3 will become negative. In other words we have the implied constraint

$$x_2 \leq 5$$

Now we can do the same thing with the second equation. Again omitting x_1 (because we know it's going to stay zero), we have

$$2x_2 + x_4 = 5$$

or

$$x_4 = 5 - 2x_2 = \frac{\frac{5}{2} - x_2}{\frac{1}{2}} \qquad (4\text{-}5)$$

which tells us that

$$x_2 \leq \tfrac{5}{2}$$

Thus, as x_2 increases, all the other nonbasic variables (in this case x_1) will remain zero. All the basic variables (in this case x_3 and x_4) will change. Eventually one of them (in this case x_4) will become zero, and this puts a limit on how far x_2 can increase. In this case x_2 can go up to $\frac{5}{2}$, and, more importantly, x_4 will be the variable that goes to zero and becomes the new nonbasic variable. This tells us where the new pivot will be in our Gaussian elimination procedure. Specifically it will occur in the x_2 column and in the row where the 1 now appears in the x_4 column, that is, in the second row.

There is a shortcut for all the arithmetic we have just done, and it is indicated in Figure 4-10, where we have copied the equations (4-2). First we have circled the x_2 column, since this is where we have decided to work (that is, x_2 will be the incoming variable). Then to the right we have written the ratios b_i/a_{ij}, where b_i is the requirement (or number on the right-hand side) in the ith row, and a_{ij} is the coefficient of x_j, the jth column being in this case the column

x_1	x_2	x_3	x_4	b	b_i/a_{ij}	
2	1	1	0	5	5	The first tableau
1	2	0	1	5	$\frac{5}{2}$	
-2	-3	0	0	$-z$		
$\frac{3}{2}$	0	1	$-\frac{1}{2}$	$\frac{5}{2}$	$\frac{5}{3}$	The second tableau
$\frac{1}{2}$	1	0	$\frac{1}{2}$	$\frac{5}{2}$	5	
$-\frac{1}{2}$	0	0	$\frac{3}{2}$	$-z+7\frac{1}{2}$		
1	0	$\frac{2}{3}$	$-\frac{1}{3}$	$\frac{5}{3}$		The third tableau
0	1	$-\frac{1}{3}$	$\frac{2}{3}$	$\frac{5}{3}$		
0	0	$\frac{1}{3}$	$\frac{4}{3}$	$-z+8\frac{1}{3}$		

FIGURE 4-11
Solving the problem (4-2) by the simplex algorithm.

we have circled. These ratios are 5 and $\frac{5}{2}$, and you will notice they are the same as the key numbers in (4-4) and (4-5). That is, all we really did in (4-4) was to show that $x_2 \leq 5$ and in (4-5) to show that $x_2 \leq \frac{5}{2}$. We have accomplished the same thing on the right of Figure 4-10. Hence we know that x_2 will increase to $\frac{5}{2}$ and that x_4 will be the variable to become nonbasic.

We have copied all this at the top of Figure 4-11 and circled this row with the smallest ratio. Then our new pivot will appear where the twice-circled element is: the 2 in the second row and second column. So the first thing we must do is divide this row by 2 and write the quotient as the second row of our second set of equations—and this is our new pivot row. We must then subtract this pivot row from the old first row to obtain the new first row. For the new third row we multiply the pivot row by 3 and add it to the old third row. Finally, therefore, we have obtained a new canonical form. Each such presentation of a canonical form is called a *tableau*, and so we have identified this set of equations as "the second tableau" [the first tableau was (4-2), our original problem]. (Incidentally, the word has a French plural: One tableau plus one tableau equals two tableaux.)

We have sneaked in a new phenomenon here. The canonical form as we have previously defined it applied only to the A-matrix (if you've forgotten what the "A-matrix" is, see Figure 4-5). But once we have our pivot, we may use it to develop zeros in the same column not only in the other rows of the

A-matrix but also in the C-vector (as we have done). We shall talk about the *canonical form of a linear programming problem*, which means, as before, that we have m columns of 1s and 0s (where m is the number of constraint equations) and, in addition, that the cost coefficients for the basic variables will also be zero.

There is a one-to-one correspondence between canonical forms and basic solutions of any linear programming problem in standard form. That is, each basic solution can be put into a unique canonical form (by Gaussian elimination); and each canonical form represents a particular basic solution (where every nonbasic variable is zero). But even when the nonbasic variables may be positive, the canonical form also represents something interesting—namely, the solution set of the original problem, expressed differently. We pointed out in connection with Figure 1-4 that we could manipulate equations and still describe the same solution set. Here the solution set (which we now call the region of feasible solutions) is a polygon (Figure 4-7 or 4-9), and this same polygon is described by the original inequalities, or by the standard form, or by any canonical form. In addition, if the nonbasic variables are allowed to go to zero, a particular canonical form represents a particular vertex of that polygon.

From this second tableau we can immediately read off the new basic solution, namely, $x_1 = x_4 = 0$ (why?), $x_2 = \frac{5}{2}$, $x_3 = \frac{5}{2}$, and $z = \frac{15}{2}$. To see what we have accomplished, refer to Figure 4-7. We started at the basic feasible solution at the origin, where $x_1 = x_2 = z = 0$. To move to a better basic solution we had to move along one of the axes; we decided to increase x_2, which meant we were going to move vertically. This meant we had to go either to the basic solution P_1, which was feasible, or to the basic solution P_2, which was infeasible. We chose, of course, P_1, which is where we are now.

It is obvious from the diagram in Figure 4-8 that we have not yet reached the optimal solution; how do we tell this from the second tableau? By looking at the C-vector. c_2 and c_3 are both zero, but the coefficients for the nonbasic variables are not zero. What happens if we increase x_4? It says that $-z$ increases (that is, z decreases) by $\frac{3}{2}$ unit for each unit of x_4; that is bad. But if we increase x_1, $-z$ will decrease by $\frac{1}{2}$ unit for each unit of increase of x_1; that is good. Therefore, we shall seek a new basic solution in which x_1 becomes a basic variable and either x_2 or x_3 becomes nonbasic. To indicate that x_1 is the incoming variable, we circle the x_1 column.

We proceed as before, taking the ratio of $\frac{5}{2}$ to $\frac{3}{2}$ in the first row, and of $\frac{5}{2}$ to $\frac{1}{2}$ in the second row, and writing these ratios to the right of the respective rows. We now select the smallest of these ratios and identify this row by circling it. The twice-circled element, $a_{11} = \frac{3}{2}$, will be our new pivot element, and so we must now divide this entire row by $\frac{3}{2}$ (or multiply it by $\frac{2}{3}$) to develop our new pivot, which is the first row in the third tableau.

We must go through the steps of the algorithm one more time. We think you can follow the development; the finished product is displayed at the bottom of Figure 4-11, and it is clear that this solution is optimal. Why? Because one of the basic feasible solutions must be optimal (Dantzig's Theorem, Section 4.3). To obtain any basic solution different from this one, either x_3 or x_4 (which are now 0) must increase, but each unit of increase of x_3 will cost us $\frac{1}{3}$ of a dollar, and each unit increase of x_4 will cost us $\frac{4}{3}$ of a dollar. Therefore, any other basic solution must be worse; so this solution must be optimal. The optimality can also be seen on Figure 4-8 where we have the same solution.

4.5 THE SIMPLEX ALGORITHM

We have now completed the development of our algorithm; it has a name—the *simplex algorithm*—and it was invented by George Dantzig in 1947 while he was working for the U.S. Air Force. It is really rather simple (although the word "simplex" doesn't mean that—it's a technical term in matrix algebra). But like long division it looks rather complicated if you try to write it down all at once. We will therefore present it in a slightly different form, in which some of the details are deferred to appendices called *subroutines*. In the same way, we could have made the long-division algorithm (Problem 1-2) more compact by stating "Step 1. Adjust the decimal point by subroutine A" and then going directly to step 8.

The Simplex Algorithm

Step 1 If the problem is not in standard form, put it in standard form by subroutine A (below). Go to step 2.

Step 2 If the A-matrix is not in canonical form, put it in canonical form by subroutine B (Section 7.1). Go to step 3.

Step 3 If the canonical form is not feasible, put it in feasible canonical form by subroutine C (Section 7.1). Go to step 4.

Step 4 If the C-vector is not in canonical form, put it in canonical form by subroutine D (Section 7.1). Go to step 5.

Step 5 Examine the coefficients of the nonbasic variables in the cost vector. If some of them are negative, go to step 6; if none are negative but some are zero, go to step 7; if all are positive, go to step 8.

Step 6 Choose the smallest (most negative) cost coefficient. Circle its column. Perform an iteration by subroutine E (below) so that this variable becomes basic. Go to step 5.

Step 7 An optimal solution has been found, but there exist alternate-optimal solutions. If you are interested in finding them, record this solution and go to subroutine F (Section 4.7); if not, go to step 8.

Step 8 The solution is optimal. Check the solution (all solutions if you have been through subroutine F) in the original problem to make sure it satisfies all constraints and yields the proper cost. Stop.

In case you've forgotten what some of the words in the first four steps mean, we define them again here.

Step 1 The problem is in standard form if and only if all constraints are in the form of equations, all variables are constrained to be nonnegative, and the objective function is to be minimized.

Step 2 The A-matrix consists of the left-hand sides of all the constraint equations (the activity coefficients). If this matrix has m rows, then it is in canonical form if and only if m of the columns constitute a permutation matrix; that is, each of these m columns must have one 1 and $m - 1$ zeros, and each of these 1s must be in a different row.

Step 3 The B-vector consists of the numbers on the right-hand sides of the constraint equations (the requirements). The canonical form is feasible if and only if every number in this vector is nonnegative.

Step 4 The C-vector consists of the numbers below the A-matrix (the cost coefficients); it is in canonical form if and only if it contains zeros under each of the m basic columns.

Subroutines B, C, and D are not needed for problems like the product-mix problem; as you noticed, when we put this problem into standard form it was already in feasible canonical form. Some of these subroutines are a little messy, and we will defer them to Section 7.1. We present here subroutines A and E, which we have already developed informally; subroutine F for alternate-optimal solutions is in Section 4.7.

Subroutine A. For putting a linear programming problem in standard form

Step A1 If the stated problem is a maximization, multiply the objective function by -1. Go to step A2.

Step A2 For each inequality constraint, convert to an equation by adding a slack variable. Go to step A3.

Step A3 For each variable in the problem which is permitted to be negative, convert to a nonnegative variable by one of the following three tricks: (*a*) If the variable x_j must be nonpositive, substitute a new nonnegative variable $x_j^* = -x_j$; (*b*) if it is possible to place a lower bound on the variable x_j—that is, if you are quite sure that x_j will be greater than $-d_j$—substitute a new nonnegative variable $x_j^* = x_j + d_j$; (*c*) otherwise, substitute two new nonnegative variables x_j^+ and x_j^-, such that $x_j = x_j^+ - x_j^-$. Go to step 2 of the simplex algorithm. (A fourth trick is explored in Problem 4-23.)

Steps A1 and A2 we have already discussed in Section 4.2. To exemplify step A3, consider an investment analysis, which is a typical application of linear programming. Our problem will be to select a set of x's, each of which represents the number of units (dollars, shares) which we will invest in a particular security (or project, or whatever). We wish to maximize our expected profits subject to appropriate constraints. Now suppose one of our options was to "sell short" a certain stock (which means selling shares we do not own in anticipation that its price will decrease); we would in effect have a negative number of shares of that stock in our portfolio. Assume that x_3 represents the number of shares of Amalgamated Widget. If we are certain we are not going to buy any of this stock, but we might sell it short, we could substitute the variable x_3^* to represent the number of shares of this stock which we would be short; this is part (*a*) of step A3. Or we might be considering both buying this stock and selling it short. If we were sure that we did not want to sell short more than 10,000 shares, then we could define the variable x_3^* to be 10,000 more than the number of shares we owned. If we really bought 8000 shares, x_3^* would be 18,000; if we sold short 8000 shares, x_3^* would be 2000. This is part (*b*) of step A3. Usually one of these tricks can be applied in practice; if not, the third trick is always applicable. x_3^+ is the number of shares we buy; x_3^- is the number we sell short. We will not want to do both. If x_j is really going to come out positive, then $x_j = x_j^+$, and $x_j^- = 0$. If x_j is really going to come out negative, then $x_j = -x_j^-$ and $x_j^+ = 0$.

Subroutine E. For iterating to obtain a new basic solution

Step E1 Call the circled column (the column of the incoming variable—that is, the variable which will become basic) j. Set $i = 0$. Go to step E2.

Step E2 Add 1 to i. If $i = m + 1$, go to step E5. If $i \leq m$, go to step E3.

Step E3 If $a_{ij} \leq 0$, go to step E2. If $a_{ij} > 0$, go to step E4.

Step E4 Compute the ratio b_i/a_{ij}, and write it to the right of the b column in the ith row. Go to step E2.

Step E5 If there are no numbers to the right of column b, the problem is unbounded. Stop. Otherwise, select the smallest of the ratios which you have written to the right of the b column. Circle this row, which is the pivot row; the twice-circled element is the new pivot element. Go to step E6.

Step E6 Perform Gaussian elimination, with the circled column (column j) the incoming variable, and the basic variable whose 1 is in the circled row the outgoing variable: Divide the circled row by the twice-circled element to form the pivot row of the new matrix. For each of the remaining m rows of the old matrix (including the cost row), develop a 0 in column j. To do this for the kth row, copy the old kth row on a work sheet; write below it the product of the new pivot row by the element a_{kj} (or c_j, if $k = m + 1$—that is, if you are in the cost row); subtract; copy the difference as the corresponding row of the new matrix. When you have developed $m + 1$ rows in the new matrix (that is, including the cost row), go to step 5 of the simplex algorithm.

You will notice that before we divide by a_{ij} in step E4, we check in step E3 to make sure it is not zero. We also do not bother with the division step if a_{ij} is negative, for reasons explained later in this chapter. The first part of step E5 refers to certain types of insoluble problems which we will worry about later. However, before getting involved with such esoteric matters, we will further examine the product-mix problem. We wish to point out how far you have come. With the algorithm as we have already developed it, you can solve any problem of the product-mix type and therefore any linear programming problem which requires maximization of a profit subject to constraints of the less-than-or-equal-to type. Our example problem was so small as to be only a toy; it had only two structural variables and two constraints. But this algorithm will work perfectly well with a real problem having, say, 100 variables and 40 constraints. Of course you would not like to do such a problem by hand. You will notice that our numbers went from integers to halves to thirds. Real problems have a nasty habit of getting involved in very messy fractions very quickly. It is no coincidence that the development of linear programming as an important practical tool was contemporaneous with the development of electronic digital computers as readily available devices; both developments took place in the 1950s.

4.6 ADDITIONAL CONSTRAINTS

Consider the following linear programming problem:

Maximize: $z = 2x_1 + 3x_2$

Subject to:

$$\begin{aligned} 2x_1 + x_2 &\leq 5 \\ x_1 + x_2 &\leq 3 \\ x_1 + 2x_2 &\leq 5 \end{aligned} \qquad (4\text{-}6)$$

This is the same problem that we had before, with the addition of a new constraint: We have only 3 pounds of copper. We now assert an important principle, which is obvious if you stop to think about it: Adding a constraint can never help, and it might hurt. In terms of linear programming, this means that the new maximum value of z which we obtain will certainly not be bigger than $8\frac{1}{3}$, and it may be smaller. We have redrawn Figure 4-8 as Figure 4-12. Our solution point was at the vertex $x_1 = x_2 = \frac{5}{3}$, but the new constraint, $x_1 + x_2 \leq 3$, cuts off this vertex (that is, makes it infeasible); in this case, therefore, z will be less than $8\frac{1}{3}$.

We ask you in Problem 4-1 to solve the problem (4-6) by the simplex algorithm and also to draw the corresponding diagram. If you do the problem correctly, you will notice that when it is put into standard form it has an additional row (for the additional constraint) and an additional column (for the additional slack variable). When you go through the first iteration, you will note that there is no difference except for the extra row and column, and you will arrive at the same vertex at the end of this iteration, namely, $x_1 = 0$, $x_2 = 2\frac{1}{2}$. Furthermore, x_1 will be selected as the new incoming variable as it was before. But

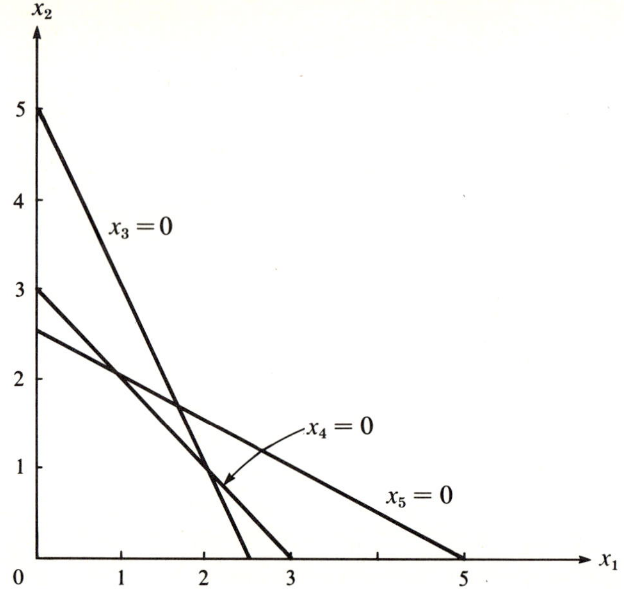

FIGURE 4-12
Graph of the problem (4-6).

at this point everything changes; x_1 cannot increase as much as before because of the new constraint, and this shows up in the algorithm by the appearance of a new smaller ratio and therefore a new row which is selected as the pivot row. We urge you to do this problem now before reading on. If you do it correctly your final tableau will be

	x_1	x_2	s_1	s_2	s_3	b
s_1	0	0	1	-3	1	1
x_1	1	0	0	2	-1	1
x_2	0	1	0	-1	1	2
	0	0	0	1	1	$-z+8$

To the left of each row we have written the name of the basic variable which had its one in that row. This optional addition to the tableau makes it easier to read and interpret; for example, "the x_2 row" (the third row) represents $x_2 = 2$ in the basic solution.

As predicted, the extra constraint has cost us a little: We make only \$8 instead of \$8.33. We achieve this new optimum by making less x_1 than before, but more x_2. Note that we must view the whole system (in this case the objective function and the constraints) in determining this optimum. If we had taken the naive, intuitive view that we were constrained away from the old optimum, so we should simply give up some of the least profitable product, we would not have obtained the best possible (optimal) solution to the new problem.

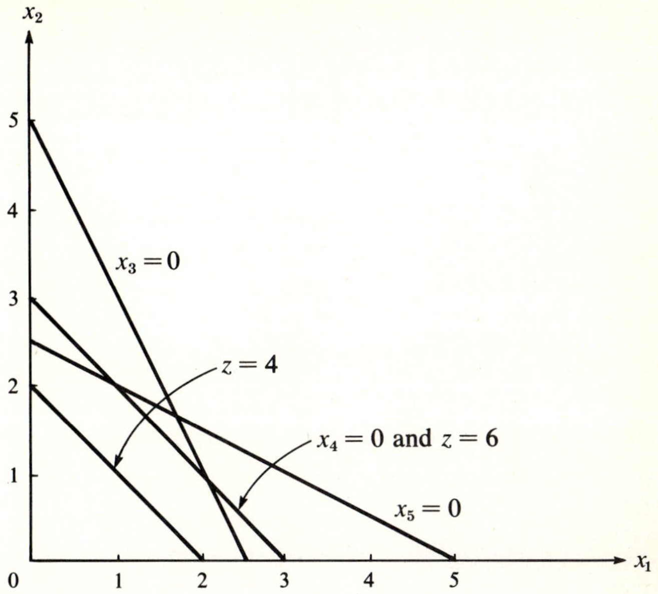

FIGURE 4-13
Graph of the problem (4-7).

4.7 ALTERNATE-OPTIMAL SOLUTIONS

Consider the following problem:

Maximize: $z = 2x_1 + 2x_2$

Subject to:

$$
\begin{aligned}
2x_1 + x_2 &\leq 5 \\
x_1 + x_2 &\leq 3 \\
x_1 + 2x_2 &\leq 5
\end{aligned}
\tag{4-7}
$$

We have graphed this problem in Figure 4-13, and present at the top of Figure 4-14 the third tableau (which you will reach after the second iteration). Please do this problem yourself before reading on to make sure that we have the right answer.

If you have done this, you will note that at the very beginning of the first iteration you are faced with a choice: The algorithm requires you to take the most negative c_j, and there are two equally negative c_j's, both of -2. Our arbitrary rule was that in case of doubt we would always pick the leftmost column. In this case, therefore, we circled the x_1-column and proceeded to reach Figure 4-14. We will indicate below what would have happened if we had circled the x_2-column instead.

In Figure 4-14 you will note that we have reached step 7 of our algorithm. There are no negative numbers left in the cost vector, and so we have reached an optimal solution. The cost coefficients c_1, c_2, and c_5, corresponding to the

	x_1	x_2	x_3	x_4	x_5	b	
x_1	1	0	1	-1	0	2	2
x_2	0	1	-1	2	0	1	
x_5	0	0	1	-3	1	1	1
	0	0	0	2	0	$-z+6$	
x_1	1	0	0	2	-1	1	
x_2	0	1	0	-1	1	2	
x_3	0	0	1	-3	1	1	
	0	0	0	2	0	$-z+6$	

FIGURE 4-14
Alternate optima.

basic variables, are 0, as they must be in a canonical form; but in addition, the cost coefficient c_3 is also 0, indicating the existence of an alternate-optimal solution. The present solution has $z = 6$, and "alternate-optimal solution" means that some other set of values of the variables also yields a profit of $z = 6$. To find out what it is requires another iteration. We do the same thing as we did before; namely, circle the nonbasic column with the smallest c_j—in this case the third column. We then proceed as before to write the ratios. Note that no ratio has been written next to the second row, because a_{23} is negative. The smallest of these ratios is 1 in the third row, and so we circle the third row. The new pivot element is a_{33}, and this is already 1; therefore, when developing the new tableau, we have to divide this entire row by 1, which means copying it. Similarly, we must develop a 0 in the cost row in the third column, and there is already a 0 there, so we also copy this row. It follows that in our new tableau we have to develop only two new rows, and this has been done at the bottom of Figure 4-14.

That the new solution is so similar to the old one is due only to the symmetry of this problem, which results from our picking numbers which made the arithmetic easy. The significant fact is that the z-lines in Figure 4-13 are parallel to one of the constraint lines; hence, when we move these cost-function lines up and to the right as far as possible, we hit two vertices at the same time. That is, the cost-function line $z = 6$ is coincident with the constraint line $x_4 = 0$—which is also the line $x_1 + x_2 = 3$. It follows that two different basic solutions, corresponding to two different vertices, give the same maximum value of profit and represent alternate-optimal solutions. Any point on this line lying between these two vertices also represents an optimal solution; for example, $x_1 = 1.4$ and $x_2 = 1.6$ is also feasible and also yields $z = 6$. This illustrates an important principle of alternate optima: When any two or more basic solutions are alternate optimal, all mixtures of these basic solutions (which constitute nonbasic solutions) are also alternate optimal. We have already seen one example of this in Section 3.3 (Figure 3-9). In problems of higher dimensionality there may be many alternate-optimal basic solutions. For example, in the problem of Figure 4-9, if the plane representing the cost function happened to be parallel to the plane $y_4 = 0$, there would be three alternate-optimal basic solutions, and all points in the triangle defined by these three vertices would be alternate optimal; if it were parallel to the plane $y_5 = 0$, there would be four different basic optimal solutions, and all points in the quadrilateral would be optimal.

Subroutine F. To find alternate-optimal solutions

Step F1 Select a nonbasic variable whose coefficient is 0. Circle its column. Iterate as before by subroutine E. Record the new basic solution. Repeat step F1 if zeros in the final tableau indicate that there are other alternate optima to be found; otherwise go to step 8 of the simplex algorithm.

Note the "other" alternate optima. The last row of Figure 4-14 indicates an alternate optimum, but it is not an "other" one. If you circled the x_5 column and iterated, obviously you would merely return to the previous solution.

4.8 TABLEAU ORGANIZATION

An interesting phenomenon appears if you choose the second column (instead of the first) in the first iteration of the problem (4-7). If you're not too tired of doing these things, it may be worth your while to explore this. You will find that after two iterations you do not arrive at the tableau on the top of Figure 4-14. You arrive instead at something very like the bottom tableau of Figure 4-14, but with the rows in different order. If you will look at Figure 4-13, you can see what has happened. In developing the top tableau of Figure 4-14, we let x_1 increase at the first iteration, which means that we went horizontally on the diagram; then on the second iteration, we went to one of the two optimal vertices—the one below and to the right. But if you let x_2 increase first, you went vertically and then over to the adjacent optimal vertex which was different.

This teaches us two things about the tableau. First, the order of the rows is immaterial. While the columns have definite identities which would vary if the columns were interchanged, the rows do not; that is, the first column continues to be associated with the first variable of the original problem, but the first row of a tableau cannot be identified with any one of the constraints of the original problem, or with any one of the rows of the original tableau, because each constraint of a succeeding tableau has in general been obtained from two

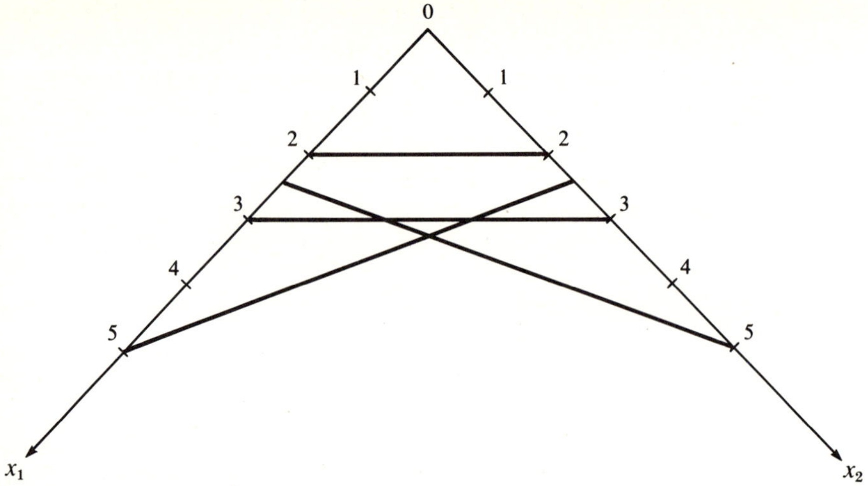

FIGURE 4-15
Figure 4-13 rotated.

of the constraints of the preceding tableau. For this reason, you may write the rows in any order you like; for example, many people prefer to put the pivot row at the top of each tableau. Of course each row of the matrix will have exactly one of the 1s in the permutation matrix, and therefore in this sense each row is identified with a particular variable. Thus in the first tableau in Figure 4-14, the first row is identified with x_1, the second row with x_2, and the third row with x_5; in the basic solution, they read $x_1 = 2$, $x_2 = 1$, and $x_5 = 1$, respectively.

As regards the choice of columns, our algorithm said that we should pick the column with the most negative c_j, but this is not really essential. In Figure 4-15 we have redrawn Figure 4-13 but have turned the world around so that the objective is to move down as far as possible. It is always possible to view a linear programming problem in this way. Now if we are at a vertex and are trying to get down as far as possible, it might be best to take the line leaving that vertex which goes down as steeply as possible; however, it might be that some other line which goes down less steeply actually goes farther, as in Figure 4-16.

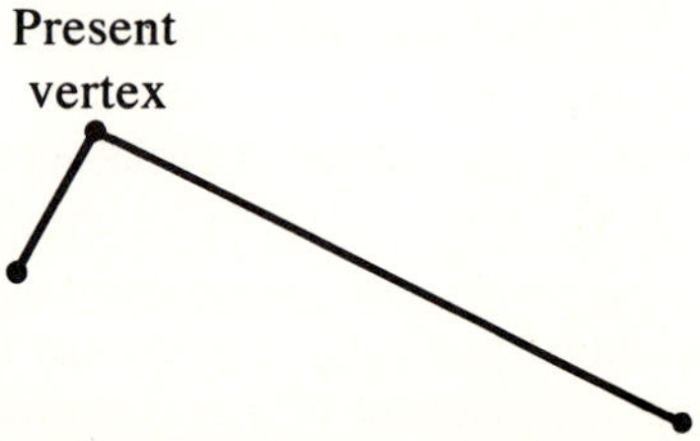

FIGURE 4-16
An unusual situation where the most negative c_j may not be best.

The distance to the first feasible vertex is measured by the ratio b_i/a_{ij} which we computed in step E4. The rate at which we go down (the slope of a line in Figure 4-15 or 4-16) is measured by c_j. The total amount we go down (the decrease in the cost function) is given by the product of these two, c_jb_i/a_{ij}. It is possible to construct a different kind of algorithm which investigates this total, but the additional complexity is probably not worthwhile. The designer of an algorithm often has such choices (more steps, each easier versus less steps, each harder); the user need not worry about them unless deeper insights into the working of the algorithm are desired.

Another way of looking at this column choice is to consider the optimization task in the simplex algorithm to be that of a bedbug who is somewhere on a large crystal chandelier. He's on a vertex, or corner, of the chandelier; he wants to get down to the bottom, and he's only allowed to go along the edges where the crystal facets meet one another. His problem is that he's very nearsighted. Being on such a vertex, one method for him to get to the bottom would be to start crawling along that edge which went down most steeply. Having arrived at the next vertex, he would reexamine the situation and again proceed along that edge which went down most steeply. When he arrived at a corner from which all edges went up, he would know he had arrived at the bottom. This bedbug algorithm might not always get him to the bottom in the shortest number of steps, but it would certainly always get him to the bottom, and it would usually be efficient. Each corner corresponds to a basic feasible solution. Going along an edge from one corner to an adjacent corner corresponds to an iteration. The steepness, down (or up) from a vertex, corresponds to the negative (or positive) magnitude of the corresponding c_j. Alternate-optimal solutions would represent an edge or facet on the bottom of the chandelier which was horizontal.

4.9 DEGENERACY

You may have wondered why we spent so much time on degeneracy in Chapter 1 and why we referred to it so mysteriously in Sections 3.3 and 3.4. It was because it has great theoretical as well as practical importance in linear programming. In this section we explore the difficulties degeneracy may lead to in the simplex algorithm and begin to understand it a little better. We shall learn much more about degeneracy in Chapters 5 and 6.

Each canonical simplex form must contain zeros in certain places, namely, in m of the $m + 1$ rows for each basic variable. A zero appearing anywhere else in the tableau represents a coincidence, which is a kind of degeneracy. We shall show below that some of these are equivalent to a previous definition of degeneracy, which implied that one equation could be derived from others. In particular, a linear programming problem is called *degenerate* if it has a zero in the B-vector (or gets one during the simplex algorithm).

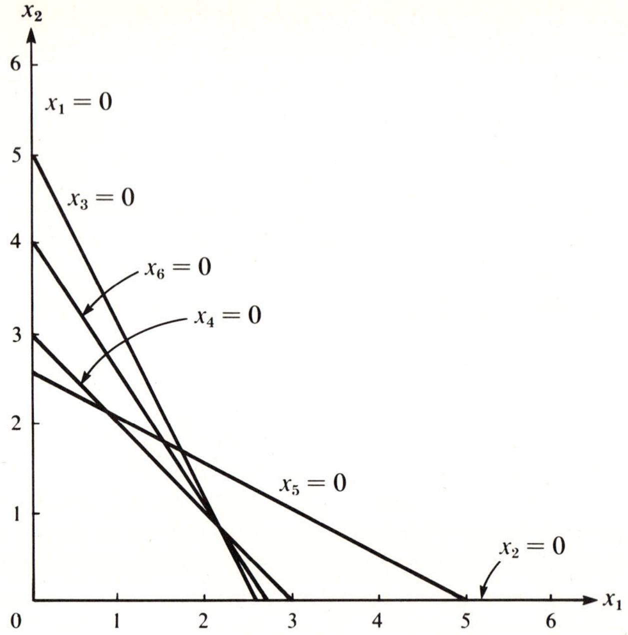

FIGURE 4-17
Graph of the degenerate problem (4-8).

Consider the problem

$$
\begin{aligned}
&\text{Maximize:} && z = 2x_1 + 3x_2 \\
&\text{Subject to:} && 2x_1 + x_2 \le 5 \\
& && x_1 + x_2 \le 3 \\
& && x_1 + 2x_2 \le 5 \\
& && 3x_1 + 2x_2 \le 8
\end{aligned}
\tag{4-8}
$$

This is identical to (4-6) with the addition of a single new constraint, but it turns out that this constraint is degenerate. The standard form is

x_3	2	1	1	0	0	0	5	$2\frac{1}{2}$	
x_4	1	1	0	1	0	0	3	3	
x_5	1	2	0	0	1	0	5	5	(4-9)
x_6	3	2	0	0	0	1	8	$2\frac{2}{3}$	
	-2	-3	0	0	0	0	$-z$		

The situation is graphed in Figure 4-17. The new constraint is the one that is identified by the label $x_6 = 0$, and this line passes through the intersection of the lines $x_3 = 0$ and $x_4 = 0$. If b_4 had been 9 instead of 8, then this constraint would have been *dominated*; that is, it would have been a superfluous constraint which could never come into play or have any significance in the problem, because any point which was feasible in accordance with the remaining constraints would automatically be feasible with respect to this one. If b_4 had been 7 instead of 8, then this constraint might have been *active*, in the sense that points that would otherwise be feasible might violate this constraint, and therefore it alone would rule out such points. The present line is just in between these; it is a superfluous constraint, but because it happens to pass through one point of the feasible region, we may arrive at it. If we do arrive at it, it will cause us trouble.

To see how this trouble develops we proceed to solve the problem. Since (4-9) represents a feasible canonical form, we skip steps 2, 3, and 4, go to step 5, and perform an iteration. In order to demonstrate degeneracy on a problem as simple as this, we ask you to take the extraordinary step of choosing the first column for this iteration instead of the second. This is unusual, but as we have indicated above, it is permissible. Note that we do not have such a choice of rows. Having chosen the first column, if we picked any row other than the first, the next canonical form would be infeasible (why?).

If you carry out the iteration in accordance with subroutine E, you will arrive at

x_1	1	$\frac{1}{2}$	$\frac{1}{2}$	0	0	0	$\frac{5}{2}$	5	
x_4	0	$\frac{1}{2}$	$-\frac{1}{2}$	1	0	0	$\frac{1}{2}$	1	
x_5	0	$\frac{3}{2}$	$-\frac{1}{2}$	0	1	0	$\frac{5}{2}$	$1\frac{2}{3}$	(4-10)
x_6	0	$\frac{1}{2}$	$-\frac{3}{2}$	0	0	1	$\frac{1}{2}$	1	
	0	-2	1	0	0	0	$-z+5$		

We are not yet finished; we must choose column 2 for the next iteration. When we get to step E5 we discover a problem. There is a tie for the smallest ratio. Of course we have an arbitrary tie-breaking rule: Pick the topmost row. However, there is an indication that degeneracy is now going to arise, because this ratio indicates the distance we are going to go along the chosen edge of the polygon to the appropriate vertex, and since two of them are at the same minimal distance, we will arrive at two vertices simultaneously (as can be seen from Figure 4-17). If we choose the x_4-row and perform the iteration, we will arrive at

x_1	1	0	1	-1	0	0	2	2	
x_2	0	1	-1	2	0	0	1		
x_5	0	0	1	-3	1	0	1	1	(4-11)
x_6	0	0	-1	-1	0	1	0		
	0	0	-1	4	0	0	$-z+7$		

Here we have degeneracy, in the form of a 0 in the b-column. It turns out in this case that, if we ignore it, nothing much will happen; when we write the ratios, we ignore the fourth row, which has the 0 in the *b*-column (because a_{43} is negative). And the next iteration brings us to

x_1	1	0	0	2	−1	0	1	
x_2	0	1	0	−1	1	0	2	
x_3	0	0	1	−3	1	0	1	(4-12)
x_6	0	0	0	−4	1	1	1	
	0	0	0	1	1	0	$-z+8$	

which is optimal. We have arrived at the same solution as before, which of course had to happen, since the additional constraint which distinguishes this example from the previous one was superfluous.

However, suppose we had written the rows in a different order, or suppose we had a different tie-breaking rule. Go back to (4-10) and see what happens when you pick the fourth row rather than the second one. The next iteration yields

x_1	1	0	2	0	0	−1	2	1	
x_4	0	0	1	1	0	−1	0	0	
x_5	0	0	4	0	1	−3	1	$\frac{1}{4}$	(4-13)
x_2	0	1	−3	0	0	2	1		
	0	0	−5	0	0	4	$-z+7$		

Now when we get to step E5 we find that our smallest ratio is 0. Remember that this ratio represented the distance we are going to move in this iteration, and the product of this ratio with the appropriate c_j represented the improvement in our objective function that we were going to achieve in this iteration. This distance and this improvement are both now 0. Again, Figure 4-17 will explain what is happening. We have arrived at the basic solution in which x_3 and x_6 are 0; in the optimal solution, x_4 and x_5 are 0. There is no way we can get from where we are to where we want to be in one iteration. We first must go from the vertex $x_3 = x_6 = 0$ to the vertex where $x_4 = x_6 = 0$. This is actually the same vertex (see Figure 4-18*a*), so it doesn't get us anywhere, and it doesn't improve our objective function. Nonetheless, we must perform this iteration in order to resolve the difficulty. Thus, in step E5, we pick the smallest ratio (namely, 0, in the second row) and perform another iteration to yield

x_1	1	0	0	−2	0	1	2	2	
x_3	0	0	1	1	0	−1	0		
x_5	0	0	0	−4	1	1	1	1	(4-14)
x_2	0	1	0	3	0	−1	1		
	0	0	0	5	0	−1	$-z+7$		

At this point a further iteration brings us to (4-12), and we are finished.

It would appear that no special provisions need be made for degeneracy; if a 0 pops up in the B-vector, and the smallest ratio turns out to be 0, then we accept the fact that for one iteration (or possibly several iterations) we get no improvement in our objective function, but eventually we should get to the right answer in any case. The difficulty, however, is more subtle. You may remember that near the end of Section 2.1 we pointed out that an effective algorithm must be guaranteed to converge—that is, come to an answer, rather than going on forever. The simplex algorithm goes from one basic solution to another, and there are only a finite number of different basic solutions, so it is guaranteed to converge as long as it never gets to the same basic solution more than once. In the absence of degeneracy it can never repeat, because the objective function improves as long as the chosen column has a strictly negative c_j and the chosen row has a strictly positive ratio (the product of these numbers measures the change in the objective function which is achieved in the iteration). And if the objective function has improved (that is, is different), then the basic solution must be different. But if that smallest ratio is 0, then the change in the objective function is 0, and we no longer have any guarantee against cycling. The algorithm cycles if it arrives at the same state repeatedly (like a stalemate in chess). In large, real problems this represents a genuine jeopardy. Such problems may be multiply degenerate; that is, they may have many zeros in the initial B-vector (we shall see in Section 4.11.2 an example of how this might arise), or many such zeros may appear simultaneously. Geometrically, this represents the intersection of several extra hyperplanes in the same vertex. Of course, you should be clever enough to recognize if you are returning to the same solution and pick a different row the next time. But remember that these large problems will be done on a computer, and the computer is not clever.

The way of getting out of this difficulty, known technically as *resolving* the degeneracy, is to add a small positive number whenever a 0 appears in the B-vector. This very small number is represented by ϵ (the Greek letter *epsilon*). Such a number is so small that there is no significant difference between, say, 1 and $1 + \epsilon$, so that if the degeneracy ever disappears, as it did in our example in going from (4-11) to (4-12), we may drop the epsilon. That is, $1 + \epsilon$ would be replaced by 1. However, mathematically the properties of even such a very small number are quite distinct from those of 0. In a computer, we might use for ϵ a number such as 10^{-12} (a millionth of a millionth). When we are working the problem by hand, we might simply write the symbol ϵ and put it in place of the 0. In either case, we are guaranteed an improvement, and be it ever so small, we cannot return to exactly the same solution.

Figure 4-18 indicates geometrically the effect of adding the epsilon. We have moved the line $x_3 = 0$ by the amount epsilon; a very small distance, insignificant in the overall problem, but sufficiently large to spread apart the degenerate vertex to a triangle. The transition, from the vertex where $x_3 = x_6 = 0$ to the vertex where $x_4 = x_6 = 0$, now represents a positive movement, and there is no way of going backward.

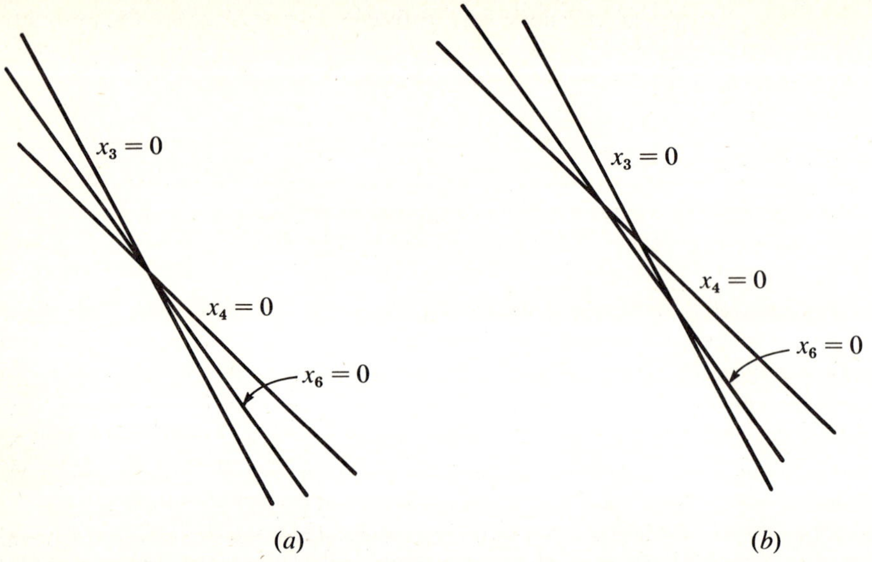

FIGURE 4-18
Expanded view of degenerate vertex. (*a*) Without ϵ; (*b*) after addition of ϵ.

Thus, if you wish, you may add an insert in step E5: "If this smallest ratio is 0, change it to epsilon." We shall need such epsilons when we develop a primal algorithm for the transportation problem since, as we shall see, degeneracy causes difficulty with that algorithm. However, we shall be particularly interested in degeneracy because of the theoretical insights it gives us into linear programming.

4.10 INFEASIBLE AND UNBOUNDED PROBLEMS

When we stated Dantzig's Theorem at the end of Section 4.3, we started "if a linear programming problem in standard form has an optimal solution . . . ." This proviso is necessary because not all linear programming problems have such solutions. In Chapter 1 we pointed out that n equations in n unknowns did not always have a unique solution and that difficulties could be of two types: Either there were no solutions or there were an infinite number of solutions; and these two situations (both degenerate) were closely related. In a somewhat similar way, difficulties with linear programming problems can arise either because there are no feasible solutions (in which case the problem is called *infeasible*) or because there are too many feasible solutions so that no matter how good a feasible solution you get there is always a better one, and no optimal feasible solution exists (in which case the problem is called *unbounded*). These two classes of problems are jointly called *insoluble problems*, and as we shall see in Chapter 5, they are related by a duality principle. Now we wish to indicate how these problems can arise, and how you will find out about them when you go through the simplex algorithm.

4.10.1 Infeasible Problems

Figure 4-19*a* represents a simple diet problem with two foodstuffs (say, meat and potatoes) on the x_1 and x_2 axes and two dietary constraints (say, protein and carbohydrate) represented by the two oblique lines. The region of feasible solutions is indicated by the shaded sides of the constraint lines and axes, and since each of the constraints is of the greater-than-or-equal-to-type, the region of feasible solutions is *unbounded*; that is, it is open and goes on to infinity in some directions. As we mentioned before, we are extending our definition of polygon to include unbounded regions such as this, whose boundaries (such as they are) are linear. There are three basic feasible solutions corresponding to the three indicated vertices of this polygon, and depending upon the relative cost of meat and potatoes, the optimal solution will occur at one of these three points (or possibly two of them, in which case there will be alternate optima).

Now suppose there is an undesirable constituent in these foodstuffs, say, arsenic, and our diet is not permitted to contain more than 15 milligrams of arsenic. The situation is illustrated in Figure 4-19*b*, where the new constraint is of the less-than-or-equal-to type, and the region of feasible solution is now the shaded region. There are still three basic feasible solutions, and again the optimal feasible solution will be at one of these three points, depending upon the relative costs of the two structural variables.

Now suppose that we discover that we are allowed to have only 5 milligrams of arsenic in the diet. The situation becomes as shown in Figure 4-19*c*, and the region of feasible solutions simply does not exist. Regardless of how much money you are willing to spend, any diet which supplies sufficient protein and carbohydrate will be infeasible because it has too much arsenic, and vice versa. The problem is insoluble because there is no feasible solution and therefore no optimal feasible solution.

A similar situation might arise in a product-mix problem. In addition to our resource constraints (only so much steel is available, and so forth), which are of the less-than-or-equal-to type, we might have constraints of the greater-than-or-equal-to type, arising, for example, because we have contracted to deliver certain minimum amounts of particular products to particular customers. In a large problem it may not be easy to determine whether or not any feasible solution exists, much less the nature of the optimal solution. Thus the difficulty may arise whenever there are constraints of the less-than-or-equal-to and greater-than-or-equal-to type in the same problem or whenever there is an equality constraint. The algorithm which we have developed works nicely whenever all the constraints are of one type, and we can skip over steps 3 and 4 of the algorithm; thus far we have learned how to handle only less-than-or-

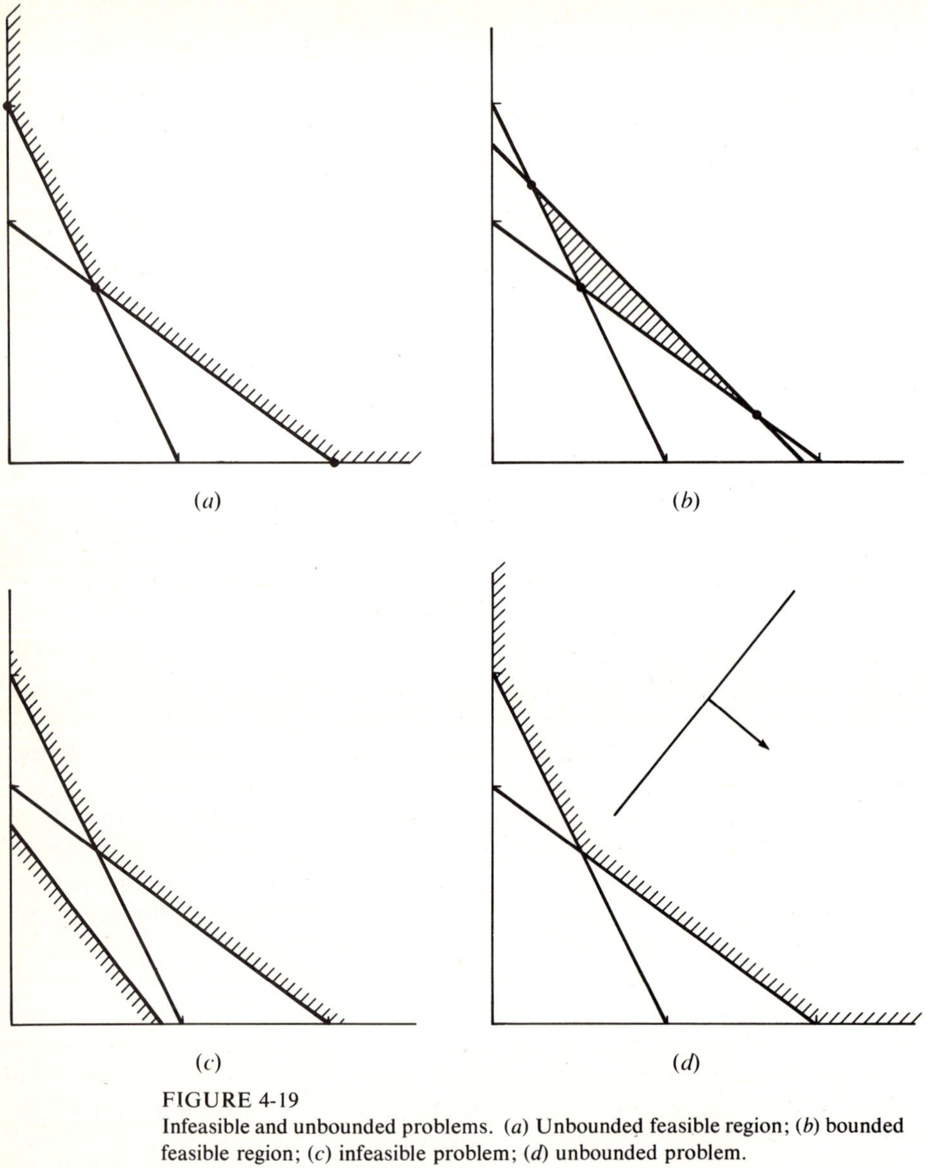

FIGURE 4-19
Infeasible and unbounded problems. (*a*) Unbounded feasible region; (*b*) bounded feasible region; (*c*) infeasible problem; (*d*) unbounded problem.

equal-to constraints, but we shall see in Section 5.1 an easy modification when all constraints are of the greater-than-or-equal-to type. Whenever both types of constraints occur in a problem, or whenever there are equality constraints in a problem, then the real difficulties (associated with lack of assurance that an optimal feasible solution exists) are reflected in difficulties with these steps in the algorithms. We shall face these difficulties and learn how to solve such problems in Section 7.1. We shall also see that the techniques developed there will allow us to solve a whole new class of problems, namely, those which are infeasible, and the question of interest is how close to feasibility we can get.

4.10.2 Unbounded Problems

Now go back to our basic two-foodstuff, two-constituent diet problem. We asserted that the optimal feasible solution would occur at one of the three vertices of the polygon of Figure 4-19*a*, depending upon the relative cost of the two foodstuffs, but it turns out that this depends on the implicit assumption that each foodstuff has a positive cost. We now ask you to consider the weird situation in which meat costs \$1 a pound and potatoes cost −\$1 a bushel; that is, the grocer gives you \$1 for each bushel of potatoes which you agree to carry away. Clearly the optimal solution is to take away an infinite amount of potatoes; the situation is graphed in Figure 4-19*d*, where the cost line is shown to go in a rather peculiar direction. The objective is to move the cost line parallel to itself in the direction of the arrow as far as possible, and obviously there is no limit on how far it can be moved.

You would probably like to know the address of this grocer, and we are happy to tell you that he doesn't exist—happy because this is equivalent to saying that such unbounded problems do not really occur. Nonetheless, negative cost coefficients can occur in the objective function of a real problem (because costs and profits may have to be considered in the same problem), and under these circumstances it is possible to make a mistake in the problem formulation so that the mathematical problem (represented by the incorrect mathematical model) is in fact unbounded. If this should occur, you will discover it in step E5 of subroutine E.

The reason why you will discover it there is quite instructive. You will remember that after the bedbug gets to a vertex, he first chooses an edge to go along (corresponding to choosing a column of the matrix to circle), and he then decides how far to go along this edge (corresponding to choosing a row to circle). The magnitude of the c_j tells him how steeply the edge goes down (or up) away from his vertex, and the ratio b_i/a_{ij} tells him how far it is to the next vertex. We pick the smallest ratio because going beyond the closest vertex in the chosen direction takes us into the infeasible region. Please note the words "in the chosen direction." Since the b_i are never negative, the ratio will be negative only when the a_{ij} is negative, and a negative ratio means a negative distance. That is, it means that the vertex in question is not in the chosen direction but in the opposite direction. That is why, in step E3, you do not compute the ratio b_i/a_{ij} if $a_{ij} \leq 0$. If every a_{ij} in the circled column is 0 or negative, then there are no vertices in the chosen direction; and since the thing that limits feasibility is the closest vertex, there is no limit on feasibility, and the solution is unbounded. You can also look at this another way, by reexamining equation (4-4), page 109. Suppose the coefficient of x_2 had been -1 instead of $+1$; then

(4-4) would read $x_3 = 5 + x_2$, and there would be no limit on how large we could make x_2.

We shall not have to look at unbounded problems except as a matter of theoretical interest. But infeasible or potentially infeasible problems will arise frequently, and we shall see them again in succeeding chapters.

4.11 PROBLEM FORMULATION

As we stated in Section 4.1.1, problem formulation is often very difficult for the beginner, because you cannot memorize a set of rules—each problem is a little different. Note the ingenuity involved, for example, in modeling the assignment problem by x's, where each x_{ij} is defined to be "the number of times the ith worker is assigned to the jth job" (which then always turns out to be either 0 or 1). In Problem 7.5-4 and at the beginning of Section 7.6 we indicate some incredibly clever variable definitions. But usually the variable definitions are straightforward. The most important aspect of problem formulation is to select and then carefully define an appropriate set of variables. These are the decision variables of the problem; they must be quantities over which the decision maker has control, and they must be of such a nature that when the decision maker has decided what these quantities are to be, then the problem will have been solved. One must always ask oneself: If I get the final answer (for example, in the form $x_1 = 3.7$, $x_2 = 23, \ldots$), does this, together with the definition of the variables, constitute a complete prescription of what action is to be taken? For example, if x_1 is "the number of pounds of beef to be purchased on Monday" or "the number of thousands of widgets to be produced per week," then the answer $x_1 = 3.7$ is meaningful. But if x_1 is something vague like "beef" or "amount of effort," then I will not know what to do with $x_1 = 3.7$, and the definition is inadequate.

You must start by writing out, in detail, on a piece of paper, "Let x_1 equal. . . ." The next words will almost always be "the number of. . . ." And the following words, which contain the substantive material of the definition, must be chosen in answer to the question, "What is my decision problem?" In the product-mix problem, we wanted to know what to manufacture, and so we "let x_1 equal the number of thousands of widgets to be produced"; in the diet problem, we "let x_1 equal the number of pounds of steak that we shall purchase each week." In some cases the necessary definitions will be more complicated; for example, in the transportation problem (and in Problem 4-3), each variable must represent a kind of interaction—between a particular origin (or cargo) and a particular destination (or hold). You will also see such interactions defined in Section 4.11.2. After you have written down the precise definition of x_1, the definitions of the other variables may usually be abbreviated: as "x_2—bushels of potatoes" or "x_3—gidgets, gross" or even by a matrix such as Figure 4-22.

But do not attempt to write equations or inequalities until you have spent the necessary few seconds to actually write down in detail precise and unambiguous definitions of the variables.

After the decision variables (which will become the structural variables of the problem) have been chosen and defined, we must choose an objective function. What is it you wish to maximize or minimize? Remember that there can be only a single objective function; it is meaningless to say, "we wish to maximize the expected profit at minimum risk." It is possible to set up a problem in which expected profit is to be maximized subject to a constraint which says that we may not exceed a certain level of risk; or it is possible to set up a problem in which we minimize the level of risk subject to a constraint which says that we must have at least a certain expected profit. In each case we have used the phrase "expected profit," which is a well-defined quantity in probability, because "to maximize profit" is something which we cannot normally do without being able to foretell the future. In each case we have also assumed that there is some appropriate numerical measure of risk; such measures exist and are described in texts on finance theory.

Finally we must formulate all the constraints. As noted above, there is sometimes a choice as to whether a particular desideratum becomes a constraint or part of the objective function. If it is to go into the objective function, it must be put into dollars or into some other terms which are commensurable with other items in the objective function. In some cases, for example, the use of an extra machine, this is easy; in other cases, for example, the number of accidents (possibly fatal) to human beings, it is difficult or impossible. In such cases it is often possible to throw this desideratum into a constraint; for example, the expected number of fatal accidents must be less than three per year.

It is desirable, but not essential, to eliminate superfluous constraints. At worst, superfluous constraints are redundant; that is, they actually duplicate others, in which case they may lead to degeneracy. If they are merely superfluous, for example, constraining something to be greater than 10 when it is already constrained to be greater than 15, then they will turn out to be inactive and will simply enlarge the problem slightly. Much more serious is the matter of omitting an essential constraint. If you do forget to write a constraint, you may end up with a solution that violates the unstated constraint. If you are lucky, this will be a nonsense solution; that is, the violation will be obvious, and it will be clear to you what new constraint must be added. You can then append this constraint, and usually get the revised solution readily by means of postoptimality analysis (Section 7.2).

An interesting story involving omission of a constraint is told by Charles Hitch, Head of Resources for Future, Inc. In the 1950s, when he was at the Rand Corporation, he had a fat colleague who wished to lose weight. After an intensive study this colleague determined that most people who go on diets do not stick to them—apparently because they get hungry. He decided to discover a diet which would have sufficient bulk to fill his stomach so that he would not become hungry. His problem formulation was very similar to our diet problem of Section 4.1.1 except that it had a different objective function and an additional constraint. The additional constraint was "not more than 1400 calories per day." The objective function was to maximize the bulk of the diet—he was sufficiently anxious to lose weight that he was willing to ignore the monetary cost. He therefore obtained data on the 437 foodstuffs that were available at the time in the Santa Monica area, including the amounts of each of the 23 dietary constituents and the caloric content and bulk of each foodstuff. He fed this data into the computer, along with the linear programming algorithm, and pressed the GO button; after a while out came the answer. As you could predict, since there were 24 constraints (the 23 dietary constituents and the calories), the solution contained 24 positive variables, of which some were slack variables (for example, he got more iron than he needed); among the 14 foodstuffs that were in the optimal diet, one was vinegar, of which, according to the computer, his diet was to include 2000 quarts per day.

Now of course the computer had not made a mistake; it had solved the problem given to it. But the problem hadn't been formulated properly; that is, he didn't really want a diet of maximum bulk. In this case he got an obviously nonsensical solution, and was able to reformulate his problem. But as Hitch points out, in large, complex problems it is not always easy to recognize vinegary solutions.

Finally, the problem formulator must decide whether or not the problem is really linear. Of course the real world is rarely exactly linear—there are usually such things as diminishing returns or quantity discounts or interactions which constitute nonlinearities—and the question is whether the linear model is a sufficiently good approximation to be useful. These things are often matters of judgment.

Bear in mind the distinction between "formulating" and "solving" a problem. Formulating as a linear programming problem means defining a set of variables, defining a linear function of those variables which is to be maximized or minimized, and defining constraints in the form of linear equations or linear inequalities in these variables. Any further steps (such as putting the problem in standard form) are part of the solution process rather than part of the formulation. We made similar comments at the end of Section 3.5, and we pointed out there that one advantage of formulating as a transportation problem is that we have a neat algorithm for solving such problems. Similarly we have a neat algorithm for solving linear programming problems. But additionally, because we know a good deal about the theory of such problems, we can tell something about the answer merely because it has been formulated; for example, we know that a certain number of the variables in any linear programming problem must be zero; and we know that, in the transportation problem, the solution will come out in integers.

	Beef trimmings	Pork trimmings	By-products	Extenders	Price per pound
Pure beef	90	0	0	10	90¢
Franks	25	45	10	20	75¢
Hot dogs	20	40	15	25	65¢
Dog food	0	0	50	50	40¢
Available	30	45	50	45	tons
Scrap value	0	0	0	$200	per ton

FIGURE 4-20
Percentage compositions of processed meat products.

4.11.1 The Sausage Problem—Product-Mix Version

To exemplify the selection and definition of variables, which we have said is the most difficult and most important part of the problem formulation, we cite two versions of the problem facing a sausage manufacturing company. To simplify the problem, we assume the company makes only four products and utilizes only four raw materials, although the real problem would be much larger.

The four products are pure-beef frankfurters, frankfurters, hot dogs, and dog food. Hot dogs are normally sold by vendors at sports events and the like; frankfurters, which are a higher quality product, are sold normally in butcher shops. The four raw materials are beef trimmings, pork trimmings, meat by-products, and extenders. The trimmings are so-called skeletal trimmings which have been scraped from the bones after the better cuts have been removed, and in the actual problem they come in various grades, depending upon the amount of fat, but we are ignoring this complication. Meat by-products consist of things such as spleens, lips, and snouts. Extenders are vegetable products such as barley and soybean; again there are great differences in the content of protein and carbohydrate in these extenders, but we also ignore this complication.

The data for the problem are presented in Figure 4-20, which shows the composition of each product in terms of the percentages of each of the raw materials of which it consists. Also shown is the price which may be obtained per pound for each product and the number of tons of each raw material available. Finally, we show that the extenders (which do not deteriorate) have some value if they are not utilized, whereas the other materials are worthless if they are not used in the very near future. We urge you to try formulating this problem before reading on.

We must first make some assumptions—or, in the real world, we must go back to the fellow who has substantive knowledge and ask him what the rules really are. In this case we shall assume complete linearity (for example, 90 pounds of beef trimmings and 10 pounds of extenders yield 100 pounds of pure-beef frankfurters) and a lack of marketing constraints; that is, we can sell as much of any of these products as we can produce. Finally we assume no complication from such things as overhead so that the numbers labeled "price" represent the actual contribution of each product to profit.

Now this problem turns out to be very similar to our product-mix problem of Section 4.1.2, and in one respect it is even a little simpler, because instead of telling you how much fat and so forth there is in each raw material, we give the actual composition of each product in terms of the amounts of raw materials that go into it. On the other hand, there are some complications. First, you must notice that the numbers in the matrix of Figure 4-20 are percentages and therefore must be divided by 100 before they are used. Next, you must transpose the matrix; that is, we have written the products as rows and the raw materials as columns; to put the problem in conventional form, you have to turn this around. Once you have done this, the definitions of variables will be pretty obvious: let x_1 equal the number of pounds of pure-beef frankfurters, x_2 equal the number of pounds of frankfurters, x_3 equal the number of pounds of hot dogs, and x_4 equal the number of pounds of dog food which we will manufacture. Then you will note that the raw materials are expressed in tons, and so you must multiply all these by 2000 to get the number of pounds that are available. Finally, you must decide what to do with the scrap-value numbers; the answer is, that you must add a term to the objective function to show that we will receive 10 cents of value for each pound of extender which we do not use. The easiest way to accomplish this is to go directly to the standard form and recognize that the amount of extender left over is simply x_8 (that is, s_4, the fourth slack variable) and introduce that into the objective function with the appropriate coefficient.

The first constraint will be $0.90x_1 + 0.25x_2 + 0.20x_3 + 0x_4 \leq 60{,}000$, which when put into standard form becomes $0.90x_1 + 0.25x_2 + 0.20x_3 + x_5 = 60{,}000$. The standard form for the entire problem will be

0.90	0.25	0.20	0	1	0	0	0	60,000
0	0.45	0.40	0	0	1	0	0	90,000
0	0.10	0.15	0.50	0	0	1	0	100,000
0.10	0.20	0.25	0.50	0	0	0	1	90,000
−0.90	−0.75	−0.65	−0.40	0	0	0	−0.10	$-z$

where the x's are in pounds and z is in dollars. We note that the first four lines are in feasible canonical form, but the C-vector is not; therefore, we must apply subroutine D (which will be given in Section 7.1) before we can actually apply

	Beef trimmings	Pork trimmings	By-products	Extenders
Pure beef	≥ 90	0	0	
Franks	≥ 25	≤ 45 ≥ 40		≤ 25
Hot dogs	≥ 20	≥ 40	≤ 20	≤ 30
Dog food			≤ 50	≤ 50

FIGURE 4-21
Revised compositions.

the simplex algorithm. Thus you cannot immediately solve this problem, unless you notice that the fourth constraint equation can be solved for x_8 to yield

$$x_8 = 90{,}000 - 0.1x_1 - 0.2x_2 - 0.25x_3 - 0.5x_4$$

Substituting this value of x_8 into the objective function yields

$$-0.90x_1 - 0.75x_2 - 0.65x_3 - 0.40x_4$$
$$- 0.10\ (90{,}000 - 0.1x_1 - 0.2x_2 - 0.25x_3 - 0.5x_4) = -z$$

which simplifies to

$$-0.89x_1 - 0.73x_2 - 0.625x_3 - 0.35x_4 = -z + 9000$$

which is in canonical form. This trick is essentially the method we shall use in subroutine D.

4.11.2 The Sausage Problem—Blending Version

Now to make the problem a little more realistic (and a lot more difficult), suppose the composition rules were given as in Figure 4-21; here, for example, frankfurters must have at least 25% beef trimmings; the amount of pork trimmings must lie between 40 and 45%; there must be less than 25% extenders; and the amount of by-products can be anything you like, provided the other constraints are satisfied. In case you want to try to formulate this problem without looking at our answer (which we urge you to do), we give you a hint: It cannot be done without using more than four variables.

	Beef trimmings	Pork trimmings	By-products	Extenders
Pure beef	x_1	x_2	x_3	x_4
Franks	x_5	x_6	x_7	x_8
Hot dogs	x_9	x_{10}	x_{11}	x_{12}
Dog food	x_{13}	x_{14}	x_{15}	x_{16}

FIGURE 4-22
Suggested variables.

The difficulty is that each of the amounts used depends upon the other amounts used. At this point you have to have a flash of genius. The trick for formulating this problem is to use a larger number of variables in which each variable represents the amount of a particular raw material used in making a particular product; for example, let x_1 equal the number of pounds of beef trimmings used in making pure-beef frankfurters. This will lead to a total of 16 variables, as indicated in Figure 4-22. We suggest it is necessary to write out the exact definition of only one of the variables, and then indicate the definition of the others by means of a matrix such as Figure 4-22. Note that at least two of these variables (namely, x_2 and x_3) will certainly be 0 and could be omitted from the problem formulation, thus leaving only 14 variables instead of 16. However, there are some advantages to using the full 16 variables, at least in the first formulation of the problem, because it will lead to a great deal of symmetry which permits one to check more easily the formulation to detect possible errors. If one is running such a problem on a computer, the available program may not be bothered by the 0 variables, and the whole thing may be just as easy with them as without them.

In terms of these variables, the formulation is easy. The constraint of availability of beef trimmings, for example, simply becomes

$$x_1 + x_5 + x_9 + x_{13} \leq 60{,}000$$

(By the way, can you see the symmetry in this expression?) Eventually you will want to know exactly how much pure-beef frankfurters will be produced, and the answer is $x_1 + x_2 + x_3 + x_4$. If you wanted to, you could define a new variable, x_{17}, which would equal this quantity. This would make the formulation a little simpler (and more degenerate) at the expense of a slightly larger problem; but the idea gives us the key as to how to take care of that at-least-90% requirement, which becomes $x_1 \geq 0.9\ (x_1 + x_2 + x_3 + x_4)$. Multiplying out and transferring all the quantities to the left-hand side of the equation (in order to be more like the standard form) we get

$$0.1x_1 - 0.9x_2 - 0.9x_3 - 0.9x_4 \geq 0$$

There will be a total of 11 such constraints, one for each of the inequalities in Figure 4-21. Thus there will be a total of 15 inequality constraints (of which 11 are immediately degenerate), plus two equality constraints for $x_2 = x_3 = 0$ (which must be included explicitly only if x_2 and x_3 are used as shown in Figure 4-22), plus the nonnegativity constraints (one for each variable). The number of variables will be 14 or 16, or 20 if you define variables such as x_{17} above (or, as we shall indicate below, perhaps only 12), plus the slack variables.

Because the constraints of Figure 4-21 are of both the greater-than-or-equal-to and less-than-or-equal-to types, the problem cannot be put directly into canonical form. Thus we shall not be able to solve this problem until we get through Section 7.1. Again, however, this has been an exercise in formulation, and the moral is that one must use a certain amount of imagination and creativity in finding the appropriate formulation for any particular problem.

One can draw some interesting conclusions about the nature of the solution by careful inspection of the data of Figure 4-20 or Figure 4-21. The problem of Figure 4-20 has four constraints, four structural variables, and four slack variables. Not more than four (and probably exactly four) of these eight variables will be positive in the final solution, and if you examine the data carefully you can make a pretty good guess as to which four they will be. For example, it seems pretty obvious that some pure-beef frankfurters will be made (because they are so profitable). Furthermore, some dog food will be made (what else are you going to do with all those by-products?). It also seems clear that one, but not both, of the remaining two products will be made (there's no other way to use pork trimmings). If all this is true, it implies that three of the four raw materials will be used up and we will have slack remaining on one of them. Which do you suppose it will be? Not the extenders, even though they are worth 10 cents. Note that each of the products uses at least as much extenders as by-products; since we have more by-products to start with, there have to be some left over after all the extenders are used up. Since there will be only one positive slack variable, it follows that all the trimmings will also be used up.

Thus, a careful examination of the numbers, bolstered by knowledge of the theory of linear programming, enables us virtually to solve this rather complicated problem by inspection. The final tableau for this problem is displayed in Problem 5-3.

For the second version of the problem, if you examine the data of Figures 4-20 and 4-21 carefully, it becomes obvious that x_{13} and x_{14} will turn out to be zero (why?). This is why we indicated above that you might use as few as 12 variables in this problem. (In fact, you could even reduce it to 11, because it is now clear that $x_{15} = x_{16}$.) Of course such guesses might be wrong, but there is not much danger in making them. You can always use postoptimality analysis

(Section 7.2) to add an additional constraint or variable to the final solution and see whether your guess was right.

4.12 IMPORTANT TERMS

The following are the most important terms defined in this chapter. You should not go on to Chapter 5 until you thoroughly understand each of them. Each is in italics where first defined.

- Linear programming problem
- Diet problem
- Product-mix problem
- Linear programming problem in standard form
- Slack variable
- Structural variable
- Requirements vector (*B*-vector, *b*-column)
- Cost vector (*C*-vector, *c*-row)
- Activity matrix (*A*-matrix)
- Feasible solution
- Optimal feasible solution
- Region of feasible solutions
- Dantzig's Theorem
- Tableaux
- Canonical form of a linear programming problem
- Simplex algorithm
- Subroutine
- Degeneracy
- Resolving (degeneracy)
- Epsilon (ϵ)
- Infeasible problem
- Unbounded problem

Problems

If you have done the problems in Chapters 2 and 3, you probably do not need to have us tell you again to make haste slowly. The important thing in doing a simplex problem by hand is not to make any arithmetic mistakes. When you circle a row, don't include the ratio (to the right of the *b* column) in the circle. You may find it convenient to put extra lines or symbols on the tableau. For example, a line to the right of the *B*-vector will keep you from confusing the ratios with the *b*'s; variable names ($x_1, x_2, \ldots, s_1, \ldots, b$) above the columns may clarify things; you may or may not wish to write variable names on the rows, as suggested in Section 4.8; and so forth. You might have noted that in Figure 4-11 we put a line between the *A*-matrix and the *C*-vector and a double line between the tableaux; but in Figure 4-14 we put a single line between tableaux and omitted the other line altogether. You should do whatever is comfortable for you. The only important things are to understand what you are doing and, incidentally, to get the right answer. If you find at any time that one of the numbers in the B-vector is negative, you probably chose the wrong row in the preceding tableau. At the end of any tableau you can write out the basic solution (including all the slack variables and the cost function) and then check it back in the first standard form. If it fails to check, go back to the preceding tableau. If you find an error, it is probably best to start all over again at that point rather than try to correct it.

4-1 (*a*) Solve the problem (4-6) in Section 4.6.
(*b*) Draw the graph of this problem and indicate the solution point.

4-2 Joe Carpenter makes chairs and tables. Each chair requires 5 board feet of lumber, one box of nails, and 2 quarts of varnish, and each sells for 5 dollars. Each table uses 6 board feet of lumber, two boxes of nails, and 1 pint of varnish, and each sells for 8 dollars. He has in stock just 74 board feet of lumber, 22 boxes of nails, and 22 quarts of varnish. What should he do? (For those of you unfamiliar with the foolish nonmetric system of measures, there are 2 pints to a quart.)
(*a*) Define an appropriate set of variables.
(*b*) Using these variables, formulate as a linear programming problem.
(*c*) Graph the problem and determine the solution from the graph.
(*d*) Write the standard form corresponding to part (*b*).
(*e*) Detach coefficients, thus yielding the first simplex tableau.
(*f*) Solve by the simplex algorithm (don't forget the hints preceding Problem 4-1).
(*g*) Check this solution.
(*h*) State in words the complete solution.
(*i*) What are the relative advantages and disadvantages of the graphical method of part (*c*) and the simplex method?

4-3 A ship has three cargo holds: forward, aft, and center. The capacity limits are

Forward	3000 tons	155,000 cubic feet
Center	4000 tons	185,000 cubic feet
Rear	2500 tons	45,000 cubic feet

Because of balance considerations, the weight in the forward hold must be exactly 30% more than the weight in the rear hold, and the weight in the center hold must be more than 30% greater than the weight in the forward hold and more than 30% greater than the weight in the rear hold, but less than 90% of their sum.

The following cargoes are offered; the shipowners may accept all or any part of each commodity:

Commodity	Amount (tons)	Volume per ton (cubic feet)	Profit per ton ($)
A	5000	60	6
B	3000	50	8
C	1000	25	5
D	1500	40	7

We wish to find the number of tons of each commodity which should be placed in each hold in order to maximize profits.

Set up the problem as a linear programming problem; that is, define a set of variables (x's), and in terms of these variables, write a profit function which is to be maximized subject to a set of linear constraints in these variables. Be sure you write all the necessary constraints, but no unnecessary ones. *Do not solve.* Do not even go to the standard form. This problem is continued in Problem 5-10.

4-4 In each of parts (*a*) and (*b*) answer in your own words, in two ways: algebraic (what happens to certain numbers, such as a_{ij}, b_i, c_j, x_j, or z) and geometric (what happens to the bedbug on the polyhedron).

(*a*) In step 6 of the simplex algorithm, we circle a particular column. What would happen if we picked a different column?

(*b*) In step E5 we circle a particular row. What would happen if we picked a different row?

In each part you should consider separately two or more cases, depending on just which kind of row (or column) you pick.

4-5 Consider the following linear programming problem:

$$\begin{array}{ll} \text{Maximize:} & z = 4x_1 + x_2 + 3x_3 \\ \text{Subject to:} & x_1 + 2x_2 + x_3 \le 5 \\ & x_1 + x_2 \le x_3 \\ & x_1 + 3x_2 + x_3 \le 8 \\ \text{and} & x_1 \ge 0 \quad x_2 \ge 0 \quad x_3 \ge 0 \end{array}$$

(*a*) Put in standard form. (*Hint:* it will be degenerate.)

(*b*) Solve.

This problem is continued in Problem 7.1-4.

4-6 Consider the following final simplex tableau, representing a solution to the problem: Maximize $z = 2x_1 + 4x_2 + x_3 + 3x_4 + 2x_5$, subject to three linear constraints, which we omit:

1	1	0	2	0	-1	0	$\frac{1}{2}$	1
0	0	0	-1	1	$\frac{1}{2}$	$\frac{1}{5}$	-1	3
-2	0	1	-1	0	5	$-\frac{3}{10}$	2	7
0	0	0	2	0	2	$\frac{1}{10}$	2	$-z+17$

(*a*) What is the solution?

(*b*) Find an alternate-optimal basic solution. What is it?

(*c*) Find an alternate-optimal nonbasic solution. What is it?

(*d*) How many variables are positive in your answer to part (*c*)? What can you say about this in the light of Dantzig's Theorem?

This problem is continued in Problem 5-4 and subsequent problems.

Drill Problems

4-7

$$\begin{array}{ll} \text{Maximize:} & x_1 + x_3 + 2x_6 \\ \text{Subject to:} & x_1 + x_4 + 6x_6 = 9 \\ & 3x_1 + x_2 - 4x_3 + 2x_6 = 2 \\ & x_1 + 2x_3 + x_5 + 2x_6 = 6 \\ \text{and} & x_1 \ge 0 \quad x_2 \ge 0 \quad x_3 \ge 0 \quad x_4 \ge 0 \quad x_5 \ge 0 \quad x_6 \ge 0 \end{array}$$

Note: This problem will give you a little practice in working with fractions. The largest denominators you should get are 12 in the third tableau and 24 in the fourth (last) tableau.

4-8 Consider the following simplex tableau:

X_1	X_2	X_3	X_4	X_5	X_6	
3	0	1	0	4	0	2
2	1	0	2	3	0	3
1	0	0	1	2	1	4
3	0	0	-1	1	0	$W+3$

(*a*) What is the present solution?

(*b*) Find a better solution.

(*c*) What is this better solution?

(*d*) Is this the best solution?

4-9 Maximize: $W = 4X + 5Y + 4Z$

Subject to:

$$\begin{aligned} X + Y \quad &\le 4 \\ Y + 2Z &\le 5 \\ 2X + 2Y + 4Z &\le 15 \end{aligned}$$

and $X \ge 0 \quad Y \ge 0 \quad Z \ge 0$

This problem is continued in Problems 7.3-3 and 7.6-1.

Some Extra Drill Problems

(*a*) Maximize: $z = 3x_1 + 5x_2$

Subject to:

$$\begin{aligned} 4x_1 + 3x_2 &\le 120 \\ 5x_1 + 7x_2 &\le 290 \end{aligned}$$

and $x_1 \ge 0 \quad x_2 \ge 0$

Solution: $x_1 = 0,\ x_2 = 40,\ z = 200$

(*b*) Maximize: $z = 3x_1 + 4x_2 + 2x_3$

Subject to:

$$\begin{aligned} 2x_1 + x_2 + x_3 &\le 48 \\ 3x_1 + 6x_2 + 3x_3 &\le 66 \\ 10x_1 + 5x_2 + 10x_3 &\le 130 \end{aligned}$$

Solution: $x_1 = 10,\ x_2 = 6,\ x_3 = 0,\ z = 54$

(*c*) Maximize: $6x_1 + 4x_2 + 5x_3$

Subject to:

$$\begin{aligned} 5x_1 + 7x_2 + 8x_3 &\le 925 \\ 6x_1 + x_2 + 15x_3 &\le 975 \\ x_1 + 3x_2 + 7x_3 &\le 545 \end{aligned}$$

and $x_1 \ge 0 \quad x_2 \ge 0 \quad x_3 \ge 0$

Solution: (159.459, 18.243, 0); $z = 1029.730$

(*d*) Maximize: $4X + 12Y + 3Z$

Subject to:

$$\begin{aligned} 0 \le X &\le 100 \\ 0 \le Y &\le 50 \\ 0 \le Z &\le 150 \\ 3X + 6Y + 2Z &\le 675 \end{aligned}$$

Solution: (25, 50, 150)

You can also make up your own maximization problems, but then you should have a calculator, because the fractions may become very messy. If you want to be certain your problem will be bounded and feasible, make all coefficients positive and all constraints $\le$; but it will be more interesting if you mix in some negative a's and c's.

Additional Formulation Problems

Some of the problems below are too large to be solved easily by hand. They may be used as formulation exercises only, but it is probably useful to have the experience of solving at least one large problem on the computer and interpreting the output. Problem 4-10 is recommended for this purpose. The computer may use some of the tricks which we will learn subsequently (Chapter 5 and Sections 7.1 through 7.3), so you may need some help from your instructor in interpreting the output, especially if you have made a mistake and the computer tries to tell you about it.

4-10 (This problem is strongly recommended when the student is ready for it if a computer is available. It has numerous subtleties, and very few students see all of them without help from the instructor.)

A company manufactures three products, A, B, and C. The maximum markets for these products are 30,000, 120,000, and 150,000 units per month, respectively. The company must manufacture at least 10,000, 20,000, and 10,000 units per month, respectively, to fulfill contractual agreements. Contributions (profit exclusive of manufacturing costs) are 75, 15, and 10 cents per unit, respectively. Each unit must go through four manufacturing stages. Pending expansion, the production capacity is 40, 250, 75, and 30 hours per shift for stages 1, 2, 3, and 4, respectively. There are 20 working days, 2 shifts per day, in the month. The number of production hours required to process 1000 units of each product through each stage is given in Table 1. The cost per hour of operating the stages is given in Table 2.

Table 1

	A	B	C
1	24	40	20
2	600	70	40
3	180	10	15
4	9	30	25

Table 2

Day	Night
\$0.10	\$0.30
0.50	0.60
0.30	0.50
0.90	0.90

(*a*) Formulate as a linear programming problem. For this purpose you may use multiple subscripts (for example, x_{ij}) and leave arithmetic implicit. Obtain approval from your instructor before proceeding.

(*b*) Formulate ready for punching into the computer. For this purpose you must have single subscripts (x_j), and the format must be as demanded by the computer.

(*c*) Run.

(*d*) Detach from the computer printout those portions which are useful; annotate them appropriately; write a brief memo for the foreman telling him what to do. Do not give him more or less flexibility than required; that is, your instructions should be specific enough to allow him to make the maximum profit, but if there are alternate-optimal solutions, he should be allowed to arrange things to suit his convenience and accommodate variables which are not in the model (what variables?).

(*e*) If you had part (*a*) to do over again, would you do it differently?

4-11 A capital budgeting problem. Twelve projects are available to be funded; each project may be funded in whole, in part, or not at all. If a project is partially funded, all earnings and cash flows are in direct proportion. The tables show, for example, that project 1 has a cash outflow the first year of \$310 (actually \$310,000, but all numbers are divided by 1000 for convenience). However, \$285 of this is capitalized (against future cash inflows) so that the earnings statement in the first year will only reflect a \$25 loss if this project is undertaken. The board of directors has decreed that for the next four years the firm must show an earnings increase of at least 6% each year over the previous year. Thus, earnings this year will be \$80 and must be at least \$84.80 next year; however, if earnings happen to be \$100 next year, then the following year's earnings must be at least \$106. If no projects are undertaken, earnings in the next four years will be \$120, \$110, \$100, and \$100, respectively. Subject to these constraints, it is desired to select those projects which will maximize the present value of all future cash flows, present value being computed by discounting all future cash by 12% per year; thus \$100 next year has present value $100/1.12 = \$89.29$; \$100 two years from now has present value $100/(1.12)^2 = \$79.72$, and so forth. The amount of capital available at this discount rate is not a limiting factor. Formulate as a linear programming problem. If you are not going to solve the problem (which requires a computer), it is probably adequate to show the coefficients in detail only for a few of the projects, with others indicated by dots.

ESTIMATED CASH FLOWS FOR PROJECTS THAT CAN BE INITIATED IN THE NEXT FOUR YEARS (THOUSANDS OF DOLLARS)

	Year											
Project	1	2	3	4	5	6	7	8	9	10	11	12
1	−310	53	68	77	83	65	55					
2	−35	−55	−40	−15	25	95	140	190	190	180	130	75
3	−15	−15	−10	25	38	67	54	32				
4		−145	28	35	42	48	40	35	28	25	17	10
5		−130	−75	−70	100	95	80	75	60	50	20	
6		−30	105	130	−15	65	−140	65	500	300		
7			−250	−145	−5	65	280	415	320	100	60	
8			−50	0	100	200	150	95				
9			−23	−10	5	20	85	65	15			
10			−80	−100	−60	20	105	120	180	180	180	180
11				−250	−50	−10	50	80	125	165	200	165
12				−300	−100	−50	65	185	425	500	450	400

ESTIMATED CONTRIBUTION TO INCOME OF THE 12 PROJECTS IN THE NEXT FOUR YEARS

	Year			
Project	1	2	3	4
1	−25	7	19	38
2	−33	−50	−40	15
3	−6	−4	−3	140
4		−70	5	30
5		−55	−10	−10
6		−10	30	80
7			−60	−35
8			−15	−5
9			−10	0
10			−30	−40
11				−100
12				−150

4-12 A portfolio selection problem. A man wishes to invest 1 million dollars in such a way as to maximize his total wealth at the end of three years, subject to the constraints that he must receive at least \$30,000, \$40,000, and \$50,000 in dividends and interest in the first, second, and third years and that at least \$200,000 must be liquid (that is, in savings accounts). No transactions are permitted during the three-year period, and dividends cannot be reinvested. To simplify, assume all given numbers are appropriately discounted. This problem is much simpler than real portfolio selection problems, where account must be taken of uncertainty.

Available securities

	Cost per share	Value after three years	Dividends per share in year 1	2	3
Savings	1	1	0.05	0.05	0.05
Bond 1	105	100	8	8	8
Bond 2	95	100	5	5	5
Stock 1	50	56	2	2.5	1.8
Stock 2	75	81	3.5	3.0	3.9
Stock 3	50	66	0	0	0
Stock 4	63	65	2.5	3.6	4.1
Stock 5	63	63	5.0	4.1	3.3

(*a*) Formulate as a linear programming problem.

(*b*) Reformulate if he wishes to maximize his total after-tax wealth and he must pay 50% on dividends and interest but only 25% on net capital gains. To make it easy, assume constraints are on before-tax dividends and interest.

(*c*) See how much of the solution you can determine by inspection. You should be able to determine at least the exact amount of one of the eight securities and have a pretty good idea of how many others will be purchased.

(*d*) If you have a computer handy, solve both problems.

(*e*) Change the dividend-and-interest constraints to $35,000, 40,000, and 45,000 and solve both problems again.

4-13 (*a*) The caterer problem. A caterer needs the following numbers of clean napkins over the next five days

M	T	W	T	F
80	60	75	40	110

There is no value to napkins after Friday. New napkins cost 30 cents apiece. Laundry, at 10 cents each, takes two days (napkins sent in Monday night are ready fo use Thursday morning, and so forth); fast laundry, at 20 cents each, takes one day (napkins sent in Monday night are ready for use Wednesday morning, and so forth). Formulate as a linear programming problem. Be sure you state all the necessary constraints; try not to include any unnecessary ones. *Hint*: You cannot do this with less than six variables; more than six is probably undesirable, but you may find 10 convenient. *Hint*: Make sure your formulation gives the proper cost (30 cents) if no napkins are required Monday to Thursday, and one on Friday.

(*b*) Reformulate as a transportation problem. *Hint*: There are four "origins" or "sources": new napkins, and dirty napkins available at the end of the first three days. There are six "destinations": the five requirements and a dummy. Give the first origin plenty of capacity—say 365, just in case nothing should ever be laundered.

(*c*) Solve the model of part (*b*). What is the solution?

(*d*) Show that this solution solves the model of part (*a*). [This is easier than solving part (*a*).]

This problem is continued in Problem 7.6-3.

4-14 Trim problem. A paper mill manufactures paper in rolls 1500 feet long and 115 inches wide. It has orders for the following rolls, all 1500 feet long: two hundred 23-inch; one hundred eighty 32-inch; two hundred thirty 44-inch. Any unused 115-inch rolls may be kept to fill future orders and are of considerable value, but rolls of smaller width (other than those required to fill the orders) are worthless. Formulate as a linear programming problem.

Hints: 1. You must exhaustively enumerate all efficient methods of cutting the 115-inch roll; for example, two 23-inch and two 32-inch (with 5 inches of waste). Inefficient methods (those which have 23 or more inches of waste) may be ignored.
2. The correct objective function is very simple in form.

4-15 A marketing-mix problem. The Cut-Up Company sells a special knife to restaurant-supply houses and hospitals. The hospitals pay a higher price, but collecting from them is very difficult. For these reasons, and because of different order sizes, the profits are different—the company estimates \$19 and \$22 per box of knives, respectively. Sales are generated by a combination of selective media advertising and calls by salespersons. There are five trained salespersons available, each of whom is available for up to 170 customer contact hours per month. The budget for media advertising is \$3500 per month. In the past, average requirements to sell a box of knives to a restaurant-supply house have been 36 minutes of sales calls and \$1 of advertising; for a hospital the corresponding figures are 18 minutes and \$2. The company is now setting its marketing plans for the next five months, and during this period it wishes to sell at least 3000 boxes of knives in each of the two sectors.

4-16 A problem involving elementary probability theory. The Upper Slobovian Air Force is planning a preemptive strike against Lower Slobovia to be coincident with a declaration of war. Their intelligence indicates that Lower Slobovia has foolishly permitted its Air Force to be so vulnerable that destruction of the airfield, or the barracks, or the fuel-storage facilities will entirely negate the Lower Slobovian Air Force. At its own single airfield, Upper Slobovia has three heavy bombers which get 2 miles per gallon, four medium bombers which get 3 miles per gallon, and five light bombers which get 4 miles per gallon. The distances from the airfield to the three targets, and the probabilities of destruction of the targets by the various types of bombers, are given in the table:

	Distance from airfield	Probability of destruction by		
		Heavy	Medium	Light
Airfield	450	0.13	0.11	0.08
Barracks	500	0.15	0.12	0.09
Fuel depot	550	0.2	0.15	0.1

If a target is not destroyed, then the mission is worthless to the Upper Slobovian planners. Air Force policy requires that a plane not take off unless it has enough fuel for the round trip to its target and return plus 100 gallons in reserve. The total amount of fuel available is 4000 gallons.

(*a*) Formulate. *Hint*: The objective is to maximize the probability of destroying at least one of the three targets. It is equivalent, and more convenient, to set up an objective function which minimizes the probability that no target will be destroyed. This function will be nonlinear, but its logarithm is linear.

(*b*) This still isn't a linear programming problem. Why?

4-17 An elementary problem. A company has two grades of inspectors, 1 and 2, the classification depending upon speed and accuracy. Grade 1 calls for an inspection speed of 20 pieces per hour with 98% accuracy and carries a wage rate of \$2.00 per hour. Grade 2 requires a speed of 15 pieces per hour with an accuracy of 95% and pays \$1.75 per hour. It is required that a minimum of 1600 pieces be inspected per 8-hour day. There are a maximum of only eight grade-1 and 10 grade-2 inspectors available for hiring this month. Each inspection error is estimated to cost \$2.00 (in terms of accepting a defective item or unnecessarily rejecting a good one). Formulate as a mathematical programming problem. *Do not solve* (although the solution is obvious—it might help you to try to guess it).

*4-18** A trucking company with $400,000 to spend on new equipment is contemplating three types of vehicles. Vehicle A has a 10-ton payload and is expected to average 35 miles per hour. It costs $8000. Vehicle B has a 20-ton payload and is expected to average 30 miles per hour. It costs $13,000. Vehicle C is a modified form of B; it carries sleeping quarters for one driver, and this reduces its capacity to 18 tons and raises the cost to $15,000 but does not affect its speed. Vehicle A requires a crew of one man, and if driven on three shifts per day could be run for an average of 18 hours per day. Vehicles B and C require a crew of two men each, but, whereas B would be driven 18 hours per day with three shifts, C could average 21 hours per day. The company has 150 drivers available each day and would find it very difficult to obtain further crews. Maintenance facilities are such that the total number of vehicles must not exceed 30.

(*a*) What should the objective be? (That is, what should be maximized or minimized?)

(*b*) Formulate as a linear programming problem.

4-19 Another scheduling problem. A company knows (as in Problem 3-3) the exact demand d_j in the jth month of the coming year. No backlogging is permitted (that is, the demand must be met). As before, there is a known cost, c_j, for storing one unit during the jth month to meet demand in a later month. In this problem, production costs do *not* vary from month to month, and production capacity is unlimited. However, there is a known cost, a, for each unit by which production is increased between month $j-1$ and month j and a known cost, b, for each unit by which production is decreased between month $j-1$ and month j (these are one-time penalties for change in production rates). Production in the last month of last year was x_0, a known constant, and inventory at the end of that month was w_0, a known constant. If a and b are high and c is low, production will be set at the annual average; if a and b are low and c is high, production will fluctuate with demand. In the intermediate case, there will be some production changes, but smoothed by producing inventory in some periods to meet demand in others. Formulate as a linear programming problem.

Hint: Use the trick of two nonnegative variables from subroutine A. Let x_j be the production in the jth month; let $x_j - x_{j-1} = y_j - z_j$, where y_j is the increase (at a cost of ay_j) and z_j is the decrease (at a cost of bz_j).

Advanced Problems

4-20 Another blending problem. A firm produces two alloys ($k = 1,2$) from three ores ($i = 1,2,3$). The ores consist of various amounts of four metals ($j = 1,2,3,4$) and some nonmetallic material. Alloy 1 must have at most 70% of metal 1, at least 20% of metal 3, and at least 20% of metal 4; alloy 2 must have between 10% and 60% of metal 2, at least 20% of metal 3, and at least 30% of metal 4. The only other specification constraints are that each alloy must be all metal. The ore compositions, costs, and availabilities are

Ore No.	Metal No. 1	2	3	4	Cost ($/ton)	Tons available
1	0.25	0.1	0.15	0.3	250	2000
2	0.15	0.2	0.3	0.05	300	2500
3	0.8	0.05	0.05	0.05	200	3000

Alloy 1 sells for $800 per ton and alloy 2 for $900 per ton. There are no marketing constraints, and production costs are independent of the product mix.

*From "Operations Research—Methods and Problems" by M. Sasieni, A. Yaspan, and L. Friedman. Copyright 1959. Used with permission of John Wiley & Sons, Inc.

(*a*) A book with a similar problem gives the following hint: "Let x_{ijk} be the amount (tons) of the *j*th metal obtained from the *i*th ore and allocated to the *k*th alloy." Formulate as a profit maximization problem using this hint.
(*b*) Show that two of the constraints are obviously redundant and should be omitted.
(*c*) Show that if ore 3 costs $1 per ton and the others $10,000 per ton, this formulation will yield the wrong answer (assuming, of course, that you have changed the appropriate cost coefficients).
(*d*) Show that adding some additional (dummy) variables with $k = 3$ will always yield the right answer.
(*e*) Show that a different formulation is much neater and has less variables. *Hint*: Use x_{ij} and y_{jk}.
(*f*) In addition to being easier, neater, and quicker and cheaper on the computer, why is the last formulation better?
(*g*) Find the optimal formulation (minimum number of variables and constraints).
(*h*) How can you prove that it is (or isn't) optimal?

4-21 Construct a linear programming problem which has alternate-optimal (nonbasic) solutions but only a single optimal basic solution. There are many ways to do this, but once you see the idea you can construct very simple examples.

4-22 Draw the graph of a nondegenerate maximization problem which has two structural variables and three constraints. Each constraint line should have a negative slope (that is, each constraint line should be of the form $a_1x_1 + a_2x_2 = a_3$, where all a's are positive). Label the constraint lines $s_1 = 0$, $s_2 = 0$, and $s_3 = 0$. There will be two vertices in the feasible region in which both structural variables are positive. Label one p_1 and the other p_2. Assume that the optimal solution is at p_1. Write out the complete tableau corresponding to the vertex p_2. It should be possible to identify every number in this tableau as being 0 or 1 or + (meaning that the number must be strictly positive) or − (meaning that the number must be strictly negative).

4-23 In several problems (for example, Problems 4-3 and 4-10), we have seen constraints in the form of equations which led to immediate degeneracies. They also led to difficulties in solving the problem, requiring a new algorithm which we have promised to explain in Section 7.1. Consider the following idea: Suppose the constraint were $x_1 + x_2 + x_3 - x_7 - x_8 - x_9 = 0$. Solve it for x_9 to yield: $x_9 = x_1 + x_2 + x_3 - x_7 - x_8$; then substitute this value for x_9 wherever it appears in other constraints and the objective function. This eliminates one row and column from the tableau, plus the degeneracy.
(*a*) Actually, this idea usually won't work. What's wrong with it?
(*b*) In some problems, this is a useful trick. When would that be?

4-24 The symbol $|x|$ is read "the absolute value of x" and means "x if $x \geq 0$, and $-x$ if $x \leq 0$"; for example, $|4| = |-4| = 4$. Convert the following to a linear programming problem:

Minimize:	$\lvert x\rvert + \lvert y\rvert + \lvert z\rvert$
Subject to:	$2x + y \leq 2$
	$3x + z = 4$

4-25 Input-output analysis. This problem goes more deeply into macroeconomics than we wanted, but it is an important application, and historically it is related to the problem Dantzig was working on when he invented the simplex method.
(*a*) Consider an economy (for example, the United States) with n industries, such as railroads, steel, and so forth, the nth being "all others." There are interdependencies here. To produce one unit of the jth commodity (say, 1 ton of steel, or 1 ton-mile of railroad freight handling) takes a_{ij} units of the ith commodity. Thus, railroads need steel, and vice versa, and each also needs some of its own output (so $a_{ii} \neq 0$). Let b_i be the total consumer demand (including households, government, and export, but not including any of the n industries) in the ith industry. And let u_j be an upper bound on the ca-

pacity of the jth industry. Formulate a linear programming model of optimization of an economy. You will have to choose an objective function; you should then define some c_j's (and assume they are known). *Hint*: If $i = 1$ is steel, $j = 1$ is also steel, so you must be careful of subscripts.

(*b*) The model above is static. Now consider a dynamic model, in which the objective is to expand the economy as much as possible. Assume that production in excess of demand in period t becomes available as additional input in period $t + 1$, provided it does not exceed a storage capacity v_i for the ith commodity (for example, $v_i = 0$ for railroad freight handling, which cannot be carried over). The capacity of the jth industry is now u_{jt} in the tth period and is a variable; it may be increased by one unit if e_{ij} units of each commodity i is used to expand it. The initial capacities, u_{j1}, are known constants, as are the initial inventories of each commodity, and the demands, b_{it}, for the ith commodity in the tth period. These demands must be met. The objective is to maximize the weighted sum of the capacities, u_{jm}, at the end of the mth (last) period.

5
DUALITY

We have seen thus far several kinds of linear programming problems, and we have developed algorithms for solving three of them: the assignment problem, the transportation problem, and the product-mix problem. The development of such algorithms—that is, the ability to solve problems—is one of the objectives of this book; another is the formulation of such problems as mathematical models. But we would now like to stress a third objective, namely, to gain an understanding which makes such algorithms seem quite reasonable. Our approach thus far has been somewhat magical; out of a clear sky we pulled a Koenig's Theorem or a canonical form and said "presto, here's the solution." From here on we shall be developing other algorithms for these and other problems, but we shall also be gaining a progressively deeper understanding as to the real significance of such methods and the nature of such problems and their solutions. Underlying most of this theory will be the concept of duality.

We initially defined duality as having something to do with the interchangeability of rows and columns. Of course it is much more than that. As we pointed out in Sections 2.3 and 3.4, and most especially in Section 3.6, if you interchange the rows and columns in a certain way, you get another problem which has the same optimal cost. One of these problems is called the primal and the other is called the dual, but the designation of which is which is somewhat arbitrary. Furthermore, we have shown that in solving one problem, we often incidentally solved the other one; that is, when we had finished solving the transportation problem by the dual algorithm, we knew the optimal values not only of the dual variables (the u_i and v_j) but also of the primal variables (the x_{ij}). And the minimum value of $z = \Sigma\Sigma c_{ij}x_{ij}$ from the primal problem was equal to the maximum value of $w = \Sigma\alpha_i u_i + \Sigma\beta_j v_j$ from the dual problem. These statements apply quite generally to linear programming problems and can be formalized into something called the *fundamental theorem of linear programming*. But there is still much more to duality: We shall see, for example, that if the primal problem is degenerate, the dual problem has alternate-optimal solutions; if the primal problem is unbounded, the dual problem is infeasible; if the primal problem has an inequality that goes the wrong way (greater-than instead of less-than), the dual problem has a variable which is constrained to be nonpositive instead of nonnegative; and so on.

We will exploit many of these properties for problem solving. For example, we may have solved a large product-mix problem, and then somebody comes along and says, "Hey, here's a new product we can make. Should we make any of it? How does this affect our optimal product mix?" Now a new product represents a new column in our matrix; the question is: Do we have to start all over to solve this new enlarged problem? In many cases we do not. Instead, we look at the dual problem, in which this addition represents a new row (instead of a new column) and therefore a new constraint. If this new constraint is satisfied by our present solution, then that present solution is optimal in the new problem and we don't have to do anything else.

All this is a taste of what is to come. We turn now to a detailed demonstration of how the application of a duality principle ("if it works for rows and columns, why not for columns and rows?") permits us to solve a new and important class of problems.

5.1 THE DUAL SIMPLEX METHOD

We have seen in Section 4.1 two basic types of linear programming problems: one, the product-mix problem, in which we were asked to maximize (a profit) subject to constraints of the less-than-or-equal-to type; and two, the diet problem, in which we were asked to minimize (a cost) subject to constraints of the greater-than-or-equal-to type. We have seen that converting a maximization to a minimization is very easy, but the conversion of one kind of constraint to the other is not quite as easy. You should also note that the kind of constraint and the kind of optimization go naturally together. Thus, to minimize subject to less-than-or-equal-to constraints would be silly; we would simply move down to the origin where all variables would be zero, and the problem would be trivial. Similarly, if we were asked to maximize subject to greater-than-or-equal-to constraints, we could move on out toward infinity, and the problem would be unbounded. Real problems are never of such silly types, but mixtures of both kinds of constraints do sometimes occur in the same problem, and these require for their solution a special trick that we shall learn in Section 7.1.

To illustrate what happens in real minimization problems, we present the following example:

$$\text{Subject to:} \quad \begin{array}{l} 2y_1 + y_2 + y_3 \geq 2 \\ y_1 + y_2 + 2y_3 \geq 3 \\ y_1 \geq 0 \quad y_2 \geq 0 \quad y_3 \geq 0 \end{array} \tag{5-1}$$

$$\text{Minimize:} \quad 5y_1 + 3y_2 + 5y_3 = w$$

If you like, you may consider this to be a diet problem in which y_1, y_2, and y_3 are meat, potatoes, and vodka, respectively, and the two constraints correspond to the minimum daily requirements of protein and vitamin E. We are asked to choose a diet which minimizes the cost. We have used the symbols y here rather than the symbols x because this problem happens to be the dual of (4-6), as we shall explain below. We have already graphed this problem in Figure 4-4.

The first thing we must do, of course, is to put the problem in standard form; this becomes

$$\begin{array}{lrrrrrr} & 2 & 1 & 1 & -1 & 0 & 2 \\ & 1 & 1 & 2 & 0 & -1 & 3 \\ \text{Minimize:} & 5 & 3 & 5 & 0 & 0 & w \end{array} \tag{5-2}$$

We do not have to multiply the objective function by -1, because it is already a minimization. But please note most carefully that the coefficients of the slack variables in this case are -1, rather than $+1$ as they were before. Please convince yourself that this is the appropriate form. In particular, for example, convince yourself that the first lines of the two forms are equivalent. Thus,

$$2y_1 + y_2 + y_3 \geq 2$$

and

$$2y_1 + y_2 + y_3 - y_4 = 2$$

represent exactly the same constraint set, provided that all y's are nonnegative.

This problem is not in canonical form, because the permutation matrix which we require must specifically have 1s and 0s, and -1s do not qualify. It is easy to convert it into canonical form by simply multiplying both sides of each constraint equation by -1:

$$\begin{array}{rrrrrr} -2 & -1 & -1 & 1 & 0 & -2 \\ -1 & -1 & -2 & 0 & 1 & -3 \\ 5 & 3 & 5 & 0 & 0 & w \end{array} \tag{5-3}$$

but this canonical form is not feasible. If we set the nonbasic variables (y_1, y_2, and y_3) to zero, the basic solution becomes $y_4 = -2$, $y_5 = -3$, $w = 0$. The solution is: Buy none of any of the foodstuffs at a total cost of 0; but unfortunately this fails to satisfy both constraints, lacking two units of protein and three units of the vitamin.

To solve this problem, we invoke a duality principle. What we did before with the cost row (the C-vector) we can now do with the requirements column (the B-vector). That is, when we had arrived at a feasible canonical form in the simplex algorithm, we had the following: a permutation matrix for the basic variables. We didn't care whether the activity coefficients for the nonbasic variables were positive, negative, or 0; all the b's were nonnegative (strictly positive in nondegenerate cases); and some of the c's were negative. We then performed iterations until all the c's were nonnegative (strictly positive for nonbasic variables unless there were alternate optima), and then we were finished. If you go through the logic by which we derived all this in Chapter 4, you will see that the logic remains valid if we substitute "row" for "column," and vice versa, in appropriate places. That is, we can proceed to iterate here, always keeping the c's nonnegative, until we eliminate all negative numbers from the b's. When this has been accomplished, we will have solved the problem.

To look at this another way, note that in the simplex algorithm we started with a problem which was feasible but nonoptimal and gradually improved by decreasing the cost, always keeping the problem feasible, until we had reached the optimal solution. Here we start with a solution which is "optimal but not feasible" and gradually improve by increasing the cost, always keeping the problem "optimal," until it becomes feasible. We put the phrase "optimal but not feasible" in quotes because it's rather peculiar. If the solution isn't feasible, then it can't really be optimal; however, we found that when all c's were nonnegative, the solution was optimal, and so it is convenient to continue to call the problem "optimal" as long as all the c's are nonnegative. It would be correct to say that the objective function is and remains better than or equal to optimal, but this is not a sufficient prescription for maintaining all c's positive, which is what we need.

The details of this algorithm, which is called the *dual simplex algorithm*, are just as you would expect. We start with the most negative number in the b-column; we could actually pick any negative number in the b-column, but generally we will approach a final solution most quickly if we pick the most negative number. In this case we choose the second row and circle it:

y_4	-2	-1	-1	1	0	-2
y_5	-1	-1	-2	0	1	-3
	5	3	5	0	0	w
	5	3	$\frac{5}{2}$			

Now of course you will take ratios in columns, where before you took them in rows. In particular you will take the ratio of c_j/a_{ij}, and in this case you will do it only for those a_{ij} in the circled row which are strictly negative (before you did it only for those a_{ij} in the circled column which were strictly positive). Since the c's in this algorithm are always positive, and we take the ratio only when the a's are negative, the ratios themselves will always be negative, and we do not have to bother writing these minus signs. Thus the ratios are written below the c-row in the preceding tableau, and we pick (as before) the smallest ratio and circle its column. By picking the smallest ratio in the simplex algorithm, we arranged that the new solution would be feasible, that is, that the b's in the next tableau would all be nonnegative. By picking the smallest ratio in the dual simplex algorithm, we guarantee that the new solution will be "optimal," that is, that the c's in the next tableau will all be nonnegative.

By this time you may be getting a feeling for what happens when we go to a dual. Not only does a row become a column, but a lot of other things change. Thus, greater-than-or-equal-to becomes less-than-or-equal-to, strictly positive becomes strictly negative, and so forth. You might have guessed, therefore, that in the dual simplex algorithm, instead of looking for the smallest ratio, we should look for the largest one. This is really what we did; but since all the ratios are negative, the largest of them is the one with the smallest negative coefficient.

Now we iterate as before. Our pivot element is the doubly circled element, namely, -2, and we must therefore divide this row by -2 to obtain our new pivot row. In all other respects the iteration is performed by the Gaussian elimination method in exactly the same way as we've done before, and the new tableau becomes

y_4	$-\frac{3}{2}$	$-\frac{1}{2}$	0	1	$-\frac{1}{2}$	$-\frac{1}{2}$
y_3	$\frac{1}{2}$	$\frac{1}{2}$	1	0	$-\frac{1}{2}$	$\frac{3}{2}$
	$\frac{5}{2}$	$\frac{1}{2}$	0	0	$\frac{5}{2}$	$w - \frac{15}{2}$
	$\frac{5}{3}$	1			5	

Now we repeat the steps of the algorithm: Choose the most negative number in the b-column (this is the $-\frac{1}{2}$ in the first row), circle this row, write the ratios below the c-row, choose the smallest ratio (this is the 1 in the second column), circle this column, iterate, and obtain

	y_1	y_2	y_3	y_4	y_5		
y_2	3	1	0	-2	1	1	
y_3	-1	0	1	1	-1	1	(5-4)
	1	0	0	1	2	$w - 8$	

We have achieved feasibility, and therefore this solution is the optimal feasible solution. It reads: $y_1 = 0$, $y_2 = 1$, $y_3 = 1$, $y_4 = s_1 = 0$, $y_5 = s_2 = 0$, $w = 8$. We now state this algorithm formally.

5.2 THE DUAL SIMPLEX ALGORITHM

The dual simplex algorithm is applicable whenever the canonical form is "optimal but not feasible"; that is, when we have a canonical form (m constraint equations in n variables, $n > m$, m of the columns each consisting of one 1 and all the rest 0s, each 1 in a different row, with 0s in the cost row under these variables), and when all the numbers in the cost row are nonnegative. In addition, we assume that some or all of the numbers in the requirements column are negative, else we would not be interested in the problem (why?). To keep the numbering consistent with our numbering for the simplex algorithm, we have omitted some steps.

Step 5. Examine the numbers in the requirements column (the B-vector). If some of them are negative, go to step 6. If none are negative, go to step 7.
Step 6. Choose the most negative number; circle its row. Perform an iteration by subroutine e (below). Go to step 5.
Step 7. An optimal solution has been found. If you are interested in alternate optima, examine the cost coefficients of the nonbasic variables, and if any zeros are found there, develop the alternate optima by subroutine F (page 119); if not, go to step 8.
Step 8. The solution is optimal and feasible. Check it in the original problem to make sure it satisfies all constraints. Stop.

Subroutine e

Step e1. Call the circled row i. Set $j = 0$. Go to step e2.
Step e2. Add 1 to j. If $j = n + 1$, go to step e5; if $j \leq n$, go to step e3.
Step e3. If $a_{ij} \geq 0$, go to step e2; if $a_{ij} < 0$ go to step e4.
Step e4. Compute the ratio c_j/a_{ij}, and write it below the c-row in the jth column. This ratio will be negative, but you should ignore that and omit the negative sign. Go to step e2.
Step e5. If there are no numbers (ratios) below the c-row, the problem is infeasible; stop. Otherwise, select the smallest ratio and circle its column. The twice-circled element will become the new pivot element. Go to step e6.
Step e6. Perform Gaussian elimination, with the circled column (column j) the incoming variable, and the basic variable whose 1 is in the circled row the outgoing variable: Divide the circled row by the twice-circled element to form the pivot row of the new matrix. For each of the remaining m rows of the old matrix (including the cost row) develop a 0 in column j. To do this for the kth row, copy the old kth row on a work sheet; write below it the product of the new pivot row by the element a_{kj} (or c_j, if $k = m + 1$—that is, if you are in the cost row); subtract; copy the difference as the corresponding row of the new matrix. When you have developed $m + 1$ rows in the new matrix (that is, including the cost row), go to step 5 of the dual simplex algorithm.

If you will compare this to the simplex algorithm, you will find that most of it is very similar, and in particular that step e6 is identical to E6. We have not mentioned the possibility of a negative number in the c-row, because we assumed that there are no negative c's when we initiate the algorithm, and if that is true and if you do it properly, none of the c's will ever become negative. You can now see why we chose "minimize" in Section 4.2 as the standard form for the simplex algorithm. As a result, in both the simplex algorithm and the dual simplex algorithm, one vector (B or C) is always nonnegative, and we eliminate negatives from the other. The final tableau is identical in both cases (both vectors are nonnegative in any optimal feasible solution), which simplifies the "postoptimality analysis" of Sections 5.4 and 7.2.

5.3 RELATIONSHIP OF DUAL AND PRIMAL PROBLEMS

The dual simplex algorithm, which we have developed above, will solve any problem like the diet problem which requires us to minimize a cost subject to constraints of the greater-than-or-equal-to type. The particular problem which we have chosen to solve happens to be the dual of one of the problems we solved in Chapter 4, which is the reason we chose y's as symbols for the variables instead of x's. It is somewhat arbitrary which of a pair of (mutually dual) problems should be called the primal and which should be called the dual; but once we have chosen a particular problem as the primal, we shall label its variables as x's and its objective function as z, and the other problem, the dual, will have its variables labeled y's and its objective function labeled w.

If you will compare the dual problem (5-1) with the primal problem (4-6) on page 115, you will see that indeed the rows of one are the columns of the other, and vice versa. Furthermore, you will see that (5-1) has greater-than-or-equal-to constraints, while (4-6) has less-than-or-equal-to, and (5-1) has the objective function to be minimized, while (4-6) has the objective function to be maximized. In both cases the variables are constrained to be nonnegative. We make all this very specific later, but at the moment we wish you to compare (5-4), the final tableau of the dual, with the final tableau of the primal [which is repeated as (5-6) on page 158] and observe a remarkable fact: The solutions to both problems are displayed in the final tableau of each. Specifically, the solution of the primal problem, which is displayed as the B-vector in (5-6), is also displayed under the slack variables in the C-vector of (5-4); and the solution of the dual problem, which is displayed as the B-vector in the final tableau of the dual problem, is displayed under the slack variables in the C-vector of the primal problem. Still more specifically, the solution of the primal problem is $x_1 = 1$, $x_2 = 2$, which you will see in the B-vector of (5-6) and which you will also see under the columns y_4 and y_5 (s_1 and s_2) in the C-vector of (5-4). And the solution of the dual problem is $y_1 = 0$, $y_2 = 1$, $y_3 = 1$, which you will see on the right-hand side of (5-4) and which you will also see under the columns x_3, x_4, and x_5 (s_1, s_2, and s_3) at the bottom of (5-6). Of course you don't really "see" $y_1 = 0$ on the right-hand side of (5-4), but by now you should recognize that it is there implicitly, because y_1 is a nonbasic variable. In the following section, we will explain why these numbers appear in both places.

Just as the values of the dual structural variables appear under the slack variables in the final tableau, so the values of the dual slack variables appear under the structural variables in the final tableau. Thus, from $c_1 = 1$, $c_2 = c_3 = 0$ in (5-4), we know that $s_1 = 1$, $s_2 = s_3 = 0$ in the other problem.

You can now appreciate one of the big advantages of duality. Whenever we are given a linear programming problem, we have our choice as to whether to solve the given problem or its dual, and we usually choose whichever is easier; either way, we get the solutions to both problems. It turns out that in large problems it is easier to solve a problem with many variables and few constraints than one with few variables and many constraints, as explained in Section 7.3. Therefore, if we are given a problem with more constraints than variables (that is, more rows than columns), we will usually choose to solve its dual. When we have finished solving it, we will look for the solution to the original problem (the optimal values of the primal variables) in the c-row of the final tableau.

In order to display formally the primal and dual problems, and thereby to demonstrate how to write the dual of any given problem, we need to introduce a new symbol: T, which stands for *transpose*, and which means "write rows for columns and columns for rows." For example, the transpose of

$$A = \begin{pmatrix} 2 & 3 & 4 \\ 5 & 6 & 7 \end{pmatrix} \text{ is } B = A^T = \begin{pmatrix} 2 & 5 \\ 3 & 6 \\ 4 & 7 \end{pmatrix} \text{ and, vice versa, } (A = B^T)$$

As a further example, (3-5) and (3-6) are transposes of one another. We will use this T-symbol not only for the A-matrix but also for B and C. The B-vector is normally a column, and its transpose, written B^T, is a row. Similarly, the C-vector is normally a row vector, and C^T is a column vector. In terms of this notation we can write the dual pair of problems in the following way:

$$\begin{array}{cc} AX \le B & A^TY \ge C^T \\ CX \le z_{\max} & B^TY \ge w_{\min} \\ X \ge 0 & Y \ge 0 \end{array} \qquad (5\text{-}5)$$

On the left we have our old friend, the product-mix problem, written in a slightly different way; that is, instead of writing "maximize $CX = z$," we have written "$CX \le z_{\max}$." Note that the old formulation is somewhat imprecise anyhow. What we mean is "choose a vector X (that is, choose a set of x's) which maximizes CX." The new formulation of (5-5) implies that any X which you choose will yield some value of $z = CX$; this z may be smaller than $z_{\max}$ the largest possible value of z, or it may be equal to z_{max}, but it surely will not be greater than z_{max}. Of course, z_{max} itself, together with the X which yields it, constitutes the solution. The advantage of writing it this way is the symmetry which you can see in (5-5); it reminds you that "max" goes with "less than or equal to," and vice versa. Note, however, that the variables are nonnegative in both problems: $X \geq 0$ and $Y \geq 0$.

If you transpose the transpose of a matrix (or vector) you get back the original matrix (or vector). It follows that if you write the dual of the dual you will get the primal. You are asked in Problem 5-1 to demonstrate this.

Inequalities (5-5) represent a standard form for the dual pair of linear programming problems. Since we have already used up the word "standard" and the synonymous word "canonical," we need a new word for this standard form, and we shall call it the *conventional primal form* (on the left), and the *conventional dual form* (on the right). Remember, though, that which of a pair of problems we call the primal, and what symbolism we use, are essentially arbitrary. The conventional primal form corresponds exactly to the product-mix problem, and the conventional dual form corresponds exactly to the diet problem. Not all problems fit automatically into one of these forms, but every problem can be converted into either of them. For example, we can convert any linear programming problem into a problem in conventional primal form by changing the objective function to a maximization (if it is not already), by converting all constraints into inequalities of the less-than-or-equal-to type (if they are not already), and by ensuring that all variables are nonnegative variables. We learned in Chapter 2 how to convert a minimization to a maximization (multiply the objective function by -1). If a constraint is of the greater-than-or-equal-to type, we have merely to multiply it by -1 to convert it to less-than-or-equal-to. This may give us a negative b, but we don't care; the requirement for nonnegative b existed only in the feasible canonical form, not in the statement of the problem in standard or conventional form. (Putting the problem in standard form is part of the solution process, not part of stating the problem.) If one of the constraints is in the form of an equation (as all constraints were in the assignment and transportation problems), we substitute two inequalities. Please convince yourself that this is a valid substitution; that is, note that the equality $\Sigma a_{ij}x_j = b_i$ represents precisely the same constraint set as the pair of inequalities $\Sigma a_{ij}x_j \geq b_i$ and $\Sigma a_{ij}x_j \leq b_i$. One of this pair of constraints goes the wrong way, but we can change that by multiplying both sides by -1. That is,

$$\sum a_{ij}x_j = b_i$$

is replaced by

$$\sum a_{ij}x_j \leq b_i$$
$$\sum(-a_{ij})x_j \leq -b_i$$

Finally, if one or more of the variables are not constrained to be nonnegative, convert them to nonnegativity by step A3 of subroutine A (page 113).

5.4 OPPORTUNITY COSTS AND SHADOW PRICES

We repeat here the final tableau of our example primal problem from page 116 for your convenience:

	x_1	x_2	s_1	s_2	s_3	b
s_1	0	0	1	−3	1	1
x_1	1	0	0	2	−1	1
x_2	0	1	0	−1	1	2
	0	0	0	1	1	$-z+8$

(5-6)

We wish to discuss particularly the numbers c_3, c_4, and c_5; that is, the numbers in the c-row under the slack variables. These numbers are 0, 1, and 1, and we have asserted that they are equal to the optimal values of y_1, y_2, and y_3, respectively. We shall now indicate why. Specifically we shall show that these numbers represent the "opportunity costs" or "shadow prices" of the primal problem and that these shadow prices are the structural variables of the dual problem. In the following discussion we shall assume no degeneracies in order to make the presentation simpler.

Consider, then, the third column of (5-6), which represents x_3 or s_1, the number of units of slack on the first resource. This resource was steel, and you remember we started with 5 tons of steel [see equations (4-6) on page 115]. Apparently we were able to use only 4 tons profitably, because (5-6) indicates that we have 1 ton left over. In order to have had any of this resource left over, this slack variable had to be a basic variable. Nonbasic variables (such as s_2 and s_3) indicate that there is no slack on those resources. If x_3 is a basic variable, then c_3 has to be zero because of the nature of the canonical form. This zero tells us precisely the opportunity cost of additional steel. In this particular case we started with 5 tons of steel, we were able to use 4 tons, and we have 1 ton left over. Now suppose somebody offered us another ton of steel; how much would it be worth to us in this particular problem? Obviously nothing; we already have more steel than we know what to do with. In other words, whenever we have slack quantities of a resource, the opportunity cost of that resource is zero, and this will be indicated by $c_j = 0$ for the corresponding slack variable.

Now look at s_3. This, you will remember, was the amount of slack on the last constraint, which was the amount of labor available. You will remember we originally had 5 man-days of labor, and (5-6) indicates that we have used it all up; that is, it indicates that $x_5 = s_3 = 0$. Apparently, if we could get more labor, we could utilize it and make more money; but how much more money? We now assert that this question is answered by $c_5 = 1$ (in the c-row under x_5). That 1 means that an additional man-day of labor will be worth \$1 to us. We now proceed to justify this assertion.

Remember our interpretation of the c's. For any nonbasic variable (the only variable for which the c's could be different from 0), if the c were, for example, −2, it meant that by bringing that variable into the product mix we would decrease our cost (or improve our profit) by 2 units of money for each unit of the product. But if the c were positive, then the opposite would be true. Thus the fact that c_5 is $+1$ in this tableau indicates that if we bring s_3 into the product mix, it will cost us $1 for each unit of s_3. Now how do you bring s_3 into the product mix? By introducing some slack on labor, that is, by using less than the 5 man-days of labor that are available. In other words, if we used only 4.9 of the 5 man-days, we would end up with $0.10 less profit. We would, of course, have to revise our program, but the optimal allocation of these resources, with only 4.9 man-days instead of 5, would yield a total profit of only $7.90 instead of a total profit of $8.00. Now the logic works exactly the same in the opposite direction; if we were able to have a negative tenth of a unit of slack, which is equivalent to what would happen if we increased the available man-days of labor from 5 to 5.1, we would be able to do $0.10 better and make $8.10. Thus the 1 under s_3 represents an opportunity cost of $1 per man-day of labor. Similarly, the 1 under s_2 represents an opportunity cost of $1 per pound for copper.

These numbers, which we have called *opportunity costs*, are called by economists *shadow prices*. They represent the value of one unit of a particular resource in this extremely small subset of the world when it is isolated from the rest of the world. In general, of course, the price of steel is more than $0 per ton, but that is all it is worth to us in this particular case because we have no use for it. On the other hand, the price of copper is less than $1 a pound, and it's worth a full $1 a pound to us. Why? Because if we got an extra pound of copper, we would not only be able to use that copper, but we would also be able to use some of the surplus steel which is now worthless to us. Thus, shadow prices may be either more or less than real prices. But regardless of the real price of copper, if we could buy copper for anything less than $1 a pound, it would be profitable for us to do so, but it would not be profitable for us to pay more than $1 a pound.

Mind you, these are *marginal* costs or prices. It's clear that we couldn't use an additional 100 pounds of copper at this price; in fact it turns out we couldn't even use a single pound. You remember that on Figure 4-12 we showed how the restriction to 3 pounds of copper forced us to reduce our profit from $8.33 to $8.00. Increasing the amount of copper corresponds to moving the copper constraint line (the line $x_4 = 0$) up and to the right (parallel to itself) (page 116). As the line moves, our solution point slides along the labor constraint line (the line $x_5 = 0$) until we run out of slack steel, and we are back at the point where our profit is $8.33. If we continue to move the copper line, it will do us no further good.

We will return to these questions of sensitivity analysis and postoptimality analysis later, but first we wish to complete our explanation of why (5-6) and (5-4) display each other's solution. We have argued that the c_j's under the slack variables represent the shadow prices of the corresponding constraints: If a slack variable is basic, then this constraint is not costing us anything, and the corresponding c_j is zero; if the slack variable is nonbasic, then the constraint is costing us something, and the corresponding c_j tells us how much. Now these shadow prices are precisely the structural variables of the dual problem.

Look again at the original formulation of the dual problem, which we repeat here for convenience:

$$\begin{aligned} \text{Minimize:} \quad & 5y_1 + 3y_2 + 5y_3 = w \\ \text{Subject to:} \quad & 2y_1 + y_2 + y_3 \geq 2 \qquad (5\text{-}1) \\ & y_1 + y_2 + 2y_3 \geq 3 \end{aligned}$$

The structural variables, y_1, y_2, and y_3, now represent the shadow prices of steel, copper, and labor, respectively. Our problem as before is to design an optimum production program, but this time we are asked to do it by juggling the shadow prices. The idea is that, if something is comparatively cheap, we will use a lot of it. Conversely, if we have used a lot of something, we are imputing to it a comparatively low price. However, this relationship does not always hold; for example, if we have a surplus of steel, the shadow price remains zero no matter how much we use.

These imputed prices (the shadow prices) are the structural variables of the dual problem. The problem should be considered from the viewpoint of a competitor who wishes to buy our resources (steel, copper, and labor); how much should the competitor offer? Let y_1, y_2, and y_3 represent the number of dollars offered per ton of steel, pound of copper, and man-day of labor, respectively. The total value of the offer is w, which the buyer wishes to minimize. But if this offer is too low, we will refuse it, choosing instead to utilize these resources to manufacture widgets or gidgets. For example, we can always make 2 dollars per batch of widgets, and unless the offer satisfies the first constraint of (5-1), it will be more profitable for us to keep our resources and use them to produce widgets; similarly, the competitor must satisfy the second constraint, lest it be more profitable to produce gidgets. Finally, the total amount the buyer offers will turn out to be \$8.00, exactly what we could obtain by utilizing the resources ourselves. More precisely, at exactly \$8.00 we are indifferent, at \$7.99 we should turn it down, but at \$8.01 we should accept. But note that this must be a package deal: We aren't permitted to accept 1 dollar for copper, turn down 0 dollars for steel, and then sell the steel to someone else; that is, the shadow prices are only valid as part of a description of the overall system.

This prospective buyer is of course fictitious. What really interest us are the marginal values of our resources. For example, if we are paying labor \$6.00 per hour, then we can afford to put them on overtime (at \$9.00 per hour) if and only if the shadow price of labor exceeds \$9.00 per hour.

Viewing the dual problem as that of a competitor opens up many new applications for linear programming. There exists a set of mathematical models, called *the theory of games*, for describing and analyzing any type of competitive situation, from parlor games through economics to warfare. The most important "games" are the so-called two-person games, in which there are only two persons (as in chess) or there are only two collections of interests (as in bridge, or football, or most wars). A further important subset is the "zero-sum" games, in which whatever one person loses, the other wins. Chess and poker are like this (assuming the house doesn't collect a percentage), although war is not. However, even war can be modeled in this way: -1 for you and $+1$ for me if I shoot down your bomber; $+1$ for you and -1 for me if it gets through my air defenses and destroys my city. We have mentioned before that the systems analyst must often choose between a realistic model that is too difficult to solve and a soluble model which may be excessively simplistic—for example, modeling a nonlinear situation by a linear programming model. Military analysts have often used two-person zero-sum game theory; the extent to which such analyses are instructive or misleading is, of course, controversial.

Now it turns out that every two-person zero-sum game is exactly equivalent to a soluble linear programming problem, and every nontrivial soluble linear programming problem is exactly equivalent to a two-person zero-sum game. The pair of dual linear programming problems represent the conflicting objectives of the two "players" of the game. Thus our pair of problems (4-6) and (5-1) would become a parlor game in which one player might call heads or tails (with probabilities related to x_1 or x_2) and the other would have three choices (y_1, y_2, or y_3); the a_{ij} are related to payoffs, depending on what each player had called. The fundamental theorem of linear programming, which says that $z_{max} = w_{min}$, becomes the fundamental theorem of game theory, which says that each player has an optimal randomized strategy which leads to w_{min} or z_{max}, respectively.

Unfortunately we cannot go farther into the subject of the theory of games in this short text. For those who are interested, we recommend "The Compleat Strategyst" by J. D. Williams (from which we quoted in "To the Student"), which is even easier to read than this book. As Williams shows in his last chapter, the simplex algorithm is the most generally useful tool for solving game-theoretic problems.

Returning now to the problem (5-1), it can be considered a primal problem, in which case y_1, y_2, and y_3 are quantities of meat, potatoes, and vodka. The first row represents the minimum requirement of protein and the second row the minimum requirement of vitamin E. In that case the dual variables are x_1 and x_2, the shadow costs of these two constraints. We ask you in Problem 5-5 to interpret the problem (4-6) as one of maximizing the weighted sum of shadow prices (or if you like, shadow costs) of these dietary constituents subject to constraints involving the actual costs of the various foodstuffs.

5.5 PROPERTIES OF DUAL PROBLEMS

In Figure 5-1 we present some of the more important properties of the dual pair of problems. The list could have been made much longer. In the first row we state that the dual of a row is a column, and we might have added that the dual of a column is a row; similarly, we could have duplicated each of the first 11 rows. Only in the twelfth row is there a lack of symmetry; the dual of an unbounded problem is always infeasible, but the dual of an infeasible problem

PRIMAL	DUAL
1. Row	Column
2. Constraint	Variable
3. $\leq$	$\geq$
4. Maximize	Minimize
5. Nonnegative variable	Nonnegative variable
6. Structural variable (primal variable)	Shadow price (dual variable)
7. Inactive constraint	Zero shadow price
8. Active constraint	Positive shadow price
9. Constraint the wrong way	Nonpositive variable
10. Equality constraint	Unconstrained variable
11. Alternate optima	Degeneracy
12a. Unbounded problem	Infeasible problem
12b. Infeasible problem	Unbounded or infeasible problem

FIGURE 5-1
Some relationships and some parallel properties in dual problems.

may be unbounded or infeasible. Furthermore, we might have indicated that, where one picks the smallest ratio in the (primal) simplex algorithm, one picks the largest ratio in the dual simplex algorithm, that there is duality between "strictly positive" and "strictly negative," and so forth. We also point out that we've been a little less than rigorous in exactly what the relationship is between the left-hand and right-hand columns on Figure 5-1. For example, rows 2, 7, 8, 9, and 10 indicate that the dual of a variable has something to do with a constraint, whereas rows 5 and 6 show the relationship between primal variables and dual variables, and row 3 shows the relationship between primal constraints and dual constraints.

The first five rows of this figure show the basic relationships between the primal and dual problems that we have indicated in equations (5-5). We have introduced the fifth row to remind you specifically that in both the primal and dual problems all variables are normally constrained to be nonnegative.

In row 9 we talk about a "constraint the wrong way." The "right" way is determined by whether the problem is a maximization or a minimization; if it is maximization the right way is less-than-or-equal-to. Suppose one of the inequality constraints goes the wrong way. We have already indicated that you can multiply the inequality by -1 and thus correct this discrepancy. This leads to a constraint with a negative requirement, and the corresponding variable in the dual problem has a negative cost. We now assure you that an equivalent is to have the dual variable constrained to be zero or negative (as in the example in Section 4.5 of the security which might be sold short but would never be bought).

In row 10 we talk about an equality constraint. At the end of the last section we showed that an equality constraint can be represented by two inequality constraints, one of which corresponds to a nonpositive dual variable and the other to a nonnegative dual variable; one of this pair of dual variables will be zero. Equivalently, the dual variable is a single variable which may be positive, negative, or zero; such a variable is said to be *unconstrained.*

In rows 7 and 8 we talk about active and inactive constraints. We have already argued that the solution point will be on a vertex; only those constraints which intersect at that vertex will be *active.* Thus, another interpretation of Dantzig's Theorem is that it tells us something about active constraints (see Problem 5-12). If you prefer, an *active constraint* is one which has no slack in the optimal feasible solution. In the case of our example problem (4-6), only the labor and copper constraints were active; we had one unit of steel left over at the end, and so this (inactive) constraint really was meaningless for the problem. If we had changed the amount of steel available (from 5 tons to, say, $4\frac{1}{2}$ or $5\frac{1}{2}$ tons), it would not have affected the answer at all. This concept can be formalized in a "theorem of complementary slackness" (Section 7.7).

The duality of alternate optima and degeneracy, indicated in row 11, is of particular interest. You remember that degeneracy is indicated by 0s in the B-vector and alternate optima by 0s in the C-vector; so you might have suspected that they would be dual to one another. In particular, we shall show in the next chapter that the assignment problem is always multiply degenerate, and this leads to enormous difficulties in attempting to solve it by the simplex algorithm, or indeed by any algorithm which works directly with the primal variables. The corresponding dual problem is usually not degenerate; it does have alternate-optimal solutions (in fact an infinite set of them), but this does not cause any significant difficulties—which is why we use dual algorithms for solving assignment problems. We shall investigate this more deeply in the next chapter.

5.6 IMPORTANT TERMS

The following important terms have been defined in this chapter. You should not go on to Chapter 6 until you thoroughly understand each of them. Each is in italics where it is first defined.

- Dual simplex algorithm
- Transpose
- Conventional primal form

- Conventional dual form
- Shadow price
- Active constraint

Problems

As before: Take your time, and check your results. If in doing the dual simplex algorithm, a negative number shows up in the C-vector, it probably means that you picked the wrong column in the preceding tableau.

5-1 (*a*) Write out in detail the conventional primal form of a linear programming problem having three variables and two constraints. The objective function will be

Maximize: $$c_1x_1 + c_2x_2 + c_3x_3 = z$$

and there will be two constraints plus the nonnegativity constraints.

(*b*) Write the dual of this problem in conventional dual form. The objective function will be

Minimize: $$b_1y_1 + b_2y_2 = w$$

and there will be three constraints plus the nonnegativity constraints.

(*c*) Multiply all four rows of this dual problem by -1, and note that this converts the dual problem into conventional primal form.

(*d*) Write the dual of this problem [of part (*c*)]

(*e*) Show that the problems of parts (*a*) and (*d*) are equivalent.

This problem is continued in Problem 5-9.

5-2 Consider the following linear programming problem:

Maximize: $$4x_1 + 3x_2 - x_3$$

Subject to:

$$\begin{aligned} x_1 + 3x_2 - x_3 &\geq 12 \\ 2x_1 - x_2 + x_3 &\leq 10 \\ x_1 + x_2 &= x_3 + 5 \\ x_1 \geq 0 \quad x_2 \geq 0 \quad & x_3 \geq 0 \end{aligned}$$

(*a*) Write the dual problem, using the rules in Figure 5-1. Call the structural variables y_1, y_2, and y_3 and the objective function w.

(*b*) Put the original problem in conventional primal form (why not conventional dual form?), using the trick at the end of Section 5.3 for the equation.

(*c*) Write the dual problem for the solution to (*b*). Call the structural variables u_1, u_2, u_3, and u_4 and the objective function v.

(*d*) Show that the answers to (*a*) and (*c*) are the same problem. What is the relationship of v and w? Of u_1 and y_1? Of u_2 and y_2? And so forth.

This problem is continued in Problem 7.2-3.

5-3 If you solve the sausage problem of Section 4.11.1, you will get the following final tableau:

x_1	x_2	x_3	x_4	x_5	x_6	x_7	x_8	
1	0	−0.02469	0	1.1111	−0.61728	0	0	11,111
0	1	0.88889	0	0	2.2222	0	0	200,000
0	0	−0.01358	0	0.11111	0.16049	1	−1.0000	31,111
0	0	0.14938	1	−0.22222	−0.76543	0	2.0000	97,778
0	0	0.05420	0	0.91111	0.80494	0	0.70000	$-z + 199{,}011$

(*a*) How much would you be willing to pay for an extra pound of each of the four constituents?

(*b*) If you take 0.9 times the shadow price of beef trimmings plus 0.1 times the shadow price of extenders, you get 89 cents (the price of pure beef frankfurters in the final formulation at the end of Section 4.11.1). Why?

(*c*) If you do the same thing with the third row of Figure 4-20 (that is, 0.2 times the shadow price of beef trimmings, and so forth), you do not get 62.5 cents (the price of hot dogs). Why?

(*d*) If you add up the four numbers in the "available" row of Figure 4-20, and multiply by 2000, you get the same number as if you add up the first four numbers in the last column of the final tableau. Why?

This problem is continued in Problem 7.2-1.

5-4 In Problem 4-6,

(*a*) What is the solution to the dual problem?

(*b*) Are there any alternate-optimal solutions? If so, find one; if not, comment on their absence.

(*c*) How many variables are positive in your answer to part (*a*)? What can you say about this in light of Dantzig's Theorem?

This problem is continued in Problems 5-11 and 7.2-2

5-5 Consider the problem (5-1) on page 151 to be a diet problem minimizing the cost of a diet of meat, potatoes, and vodka, subject to requirements on protein and vitamin E. Interpret the problem (4-6) on page 115 as the dual of this, identifying each number and symbol in the tableau in terms of its meaning.

5-6 Solve by the dual simplex method:

Minimize: $3x_1 + 4x_2 + x_3$

Subject to:

$$\begin{aligned} 2x_1 + x_2 \quad\quad &\geq 10 \\ x_1 \quad\quad + 4x_3 &\geq 20 \\ 3x_2 + x_3 &\geq 40 \\ x_1 \geq 0 \quad x_2 \geq 0 \quad x_3 &\geq 0 \end{aligned}$$

This problem is continued in Problem 7.1-2.

Drill Problems

5-7 Minimize: $z = 2x_1 + 3x_2$

Subject to:

$$\begin{aligned} x_1 + x_2 &\geq 2 \\ x_1 + 2x_2 &\geq 3 \\ x_1 \geq 0 \quad x_2 &\geq 0 \end{aligned}$$

(*a*) By the graphical method.

(*b*) By the dual simplex method.

This problem is continued in Problem 7.1-3.

5-8 Minimize: $2x_1 + 5x_2 + x_3 = w$

Subject to:

$$\begin{aligned} 2x_1 + 3x_2 + 4x_3 &\geq 22 \\ x_1 - 2x_2 + 5x_3 &\geq 21 \\ 3x_1 + x_2 - 2x_3 &\geq 1 \\ x_1 \geq 0 \quad x_2 \geq 0 \quad x_3 &\geq 0 \end{aligned}$$

Use the dual simplex algorithm. Only two iterations will be required, and the optimal values of the variables will be integers. You will be involved with fractions—quarters in the second tableau and sixteenths in the third. However, as explained in Section 7.3, you do not really have to compute all the a_{ij}'s in the third tableau once you realize that the third tableau is the last one.

5-9 (*a*) Repeat Problem 5-1*a*.

(*b*) Put in standard form (five variables, two constraints, plus nonnegativity constraints).
(*c*) Write the dual of this problem (two variables, five constraints). Call the variables y_1' and y_2' and the objective function w'. *Hint*: Note row 10 of Figure 5-1.
(*d*) Show that this problem is identical to the answer to 5-1*b*. What are the relationships of y_1 to y_1', y_2 to y_2', and w to w'?

5-10 Write the dual of Problem 4-3.

Extra Drill Problems

(*a*) Minimize: $z = 4x_1 + 6x_2 + 18x_3$

Subject to:

$$x_1 + 3x_3 \geq 3$$
$$x_2 + 2x_3 \geq 5$$
$$x_1 \geq 0 \quad x_2 \geq 0 \quad x_3 \geq 0$$

Solution: $X = (0, 3, 1)$

(*b*) Minimize: $7x_1 + 11x_2 + 9x_3$

Subject to:

$$11x_1 + 7x_2 + 8x_3 \geq 130$$
$$x_1 + x_2 + 2x_3 \geq 50$$
$$18x_1 + 10x_2 + 15x_3 \geq 195$$
$$x_1 \geq 0 \quad x_2 \geq 0 \quad x_3 \geq 0$$

Solution: $X = (0, 0, 25)$

The dual of any problem in Chapter 4 is a reasonable problem. You may also make up your own minimization problems, but see the note after "Extra Drill Problems" in Chapter 4.

Advanced Problems

5-11 (*a*) Using only the tableau presented in Problem 4-6, and without solving the dual problem, write out as much of the complete final tableau of the dual problem as you can. You should complete the *B*- and *C*-vectors in detail, as well as the five basic columns of the 5×8 *A*-matrix; it is somewhat arbitrary just where you place the 1 in each basic column, so (to be specific) make the first four rows and columns of the *A*-matrix an identity matrix. For the three nonbasic columns, use the symbol $\times$ to indicate an arbitrary number.
(*b*) Using the tableau in your answer to Problem 4-6*b*, do the same thing. To maintain comparability, put the rows (the *b*'s, and the 1s in the basic columns) in the same order as in part (*a*).
(*c*) How do you explain the fact that you have two different "final" tableaux for a problem that has no alternate-optimal solutions? Specifically, if you did solve this problem, which final tableau would you get? Why wouldn't you get the other? How can you go from one to the other? Why can't you use the same method to go to still others?

5-12 Make a simple quantitative statement about the relationship between the number of active constraints and the number of structural variables in a solution at a positive level (for example, the number of products produced). Why is this relation valid? What conditions must be met for this relation to be valid? What happens if they are not met?

5-13 (*a*) Prove that if the primal is unbounded, the dual must be infeasible. *Hint*: First prove the "weak theorem of duality"; namely, if z is any feasible solution to a problem in conventional primal form, and w is any feasible solution to its dual, then $z \leq w$.
(*b*) Prove the same result in the following way: Consider how unboundedness shows up in the simplex algorithm of Chapter 4. Associated with this column is a constraint in the dual; show that this constraint cannot be made feasible.

(*c*) What happens if you try to prove that "if a problem is infeasible its dual must be unbounded" (which is false) by an analogous method using the dual simplex algorithm?

5-14 Suppose you had a large minimization problem with constraints of both less-than-or-equal-to and greater-than-or-equal-to types, which could not be gotten into canonical form for the dual simplex algorithm because a couple of cost coefficients were negative. Conventionally this problem would be solved by the rather messy techniques of Section 7.1.

(*a*) Show how to use the dual simplex algorithm *Hint*: For each j for which $c_j < 0$, substitute for x_j.

(*b*) Why would this not be a good method if many c_j were negative?

6

THE TRANSPORTATION PROBLEM AGAIN

You now know enough about linear programming, and in particular about the theory of duality, so that you can understand a great deal more about the assignment and transportation problems than you did before. It turns out that there was a reason for everything we did in Chapters 2 and 3. We are going to explain some of these reasons. We are also going to show why these problems are always degenerate and sometimes highly degenerate and what effect this has on our algorithms; how to resolve this degeneracy and thereby construct a workable primal algorithm; why the solutions always come out in integers; how to get a new answer without completely re-solving the problem if a cost or requirement has changed; the meaning of shadow prices and why the addition of requirements to a transportation problem can sometimes actually reduce the cost; and lots more. While we are at it, we are going to develop a deep understanding of these problems—and remember that such understanding is a principal objective of this book.

6.1 REVIEW

Since we are about to undertake some advanced material in the transportation problem, you should be pretty familiar with the elementary material covered in Chapter 3. You may wish to check the important terms at the ends of Chapters 2 and 3. Most of what you have to know is summarized in Figures 6-1 through 6-5, which present all our notation, as specialized for a transportation problem with three origins and four destinations. You should understand how these matrices would be modified for different transportation problems or for an assignment problem. Most especially, you should understand clearly the distinction between primal variables (the x's, which are the quotas) and dual variables (the u's and v's, which are the amounts subtracted from rows and columns respectively); and you should know what is meant by "a set of optimal primal variables" and "a set of optimal dual variables" (either of which leads directly to—or constitutes—an optimal feasible solution). If we had followed the notation of Chapter 5, we would have called the dual variables (in Figures 6-3 and 6-5) y_1 to y_7 instead of u_1 to u_3 and v_1 to v_4, but it is convenient to have separate names for the row and column variables.

Before we go any further, we must point out a possible source of confusion in notation. We have always used the symbols m for the number of rows in a problem and n for the number of columns in a problem; but unfortunately in this case that usage is ambiguous. If we consider the cost matrix, the primal variables matrix, or the dual variables matrix, and if we call the number of rows in any one of these matrices m and the number of columns n (these will be $m = 3$ and $n = 4$ in Figures 6-1, 6-2, or 6-3), then in the constraint matrix of the primal problem (Figure 6-4), the number of rows would be $m + n$ (7 in our example), and the number of columns $m \times n$ (12 in our example). If we are not careful we will be writing equations like $m = m + n$ and $n = m \times n$, which are nonsense. The mathematician's solution to such a problem would normally be to introduce new symbols. We will keep such new symbols to a minimum, but we must have two different column symbols in equation (6-2) below, because one refers to the (four) columns of Figure 6-2 and the other to the (12) columns of Figure 6-4. In addition, we will restate Dantzig's Theorem from page 107 so that it does not use m and n at all: If a linear programming problem in standard form has an optimal feasible solution, then there will be an optimal feasible basic solution in which the number of variables which have positive values is no greater than the number of constraint equations. Please convince yourself that this statement is equivalent to the one on page 107. The variables which are not positive are, of course, equal to zero (because all variables in the standard form must be nonnegative). Most of them will be nonbasic variables; some of them may be basic variables equal to zero because of degeneracy. In the absence of degeneracy, the number of positive variables will be precisely equal to the number of constraints (in the standard form all constraints are in the form of equations).

x_{11}	x_{12}	x_{13}	x_{14}	x_1	x_2	x_3	x_4
x_{21}	x_{22}	x_{23}	x_{24}	x_5	x_6	x_7	x_8
x_{31}	x_{32}	x_{33}	x_{34}	x_9	x_{10}	x_{11}	x_{12}

(*a*) (*b*)

FIGURE 6-1.
The matrix of primal variables. (*a*) Double-subscript notation; (*b*) single-subscript notation.

	β_1	β_2	β_3	β_4
α_1	c_{11}	c_{12}	c_{13}	c_{14}
α_2	c_{21}	c_{22}	c_{23}	c_{24}
α_3	c_{31}	c_{32}	c_{33}	c_{34}

(*a*)

	β_1	β_2	β_3	β_4
α_1	c_1	c_2	c_3	c_4
α_2	c_5	c_6	c_7	c_8
α_3	c_9	c_{10}	c_{11}	c_{12}

(*b*)

FIGURE 6-2.
The matrix of costs and vectors of requirements. (*a*) Double-subscript notation; (*b*) single-subscript notation.

In Chapters 2 and 3 we developed formal algorithms for the assignment and transportation problems, each of which was a dual algorithm; that is, like the dual simplex algorithm, the solution was always "optimal" (assignments were made only where the relative cost was 0), but the solution was never feasible (in the sense that all assignments had been made and the constraints thereby satisfied) until the final solution had been reached. Looked at another way, these algorithms were dual algorithms because they manipulated the dual variables, u_i and v_j. The objective function was to maximize the weighted sum of dual variables (in the assignment problem the simple sum of dual variables), and each time we went through step 8 the sum increased, until finally it reached its maximum value at the point where the solution became feasible (that is, when all discrepancies were reduced to 0). At all times, the values of the primal variables were also displayed. The basic primal variables, namely, those which were greater than 0, were displayed explicitly in the form of quotas (stars in the

				u_1
				u_2
				u_3
	v_1	v_2	v_3	v_4

FIGURE 6-3.
The vectors of dual variables.

x_1	x_2	x_3	x_4	x_5	x_6	x_7	x_8	x_9	x_{10}	x_{11}	x_{12}	
1	1	1	1	0	0	0	0	0	0	0	0	$= \alpha_1$
0	0	0	0	1	1	1	1	0	0	0	0	$= \alpha_2$
0	0	0	0	0	0	0	0	1	1	1	1	$= \alpha_3$
1	0	0	0	1	0	0	0	1	0	0	0	$= \beta_1$
0	1	0	0	0	1	0	0	0	1	0	0	$= \beta_2$
0	0	1	0	0	0	1	0	0	0	1	0	$= \beta_3$
0	0	0	1	0	0	0	1	0	0	0	1	$= \beta_4$
c_1	c_2	c_3	c_4	c_5	c_6	c_7	c_8	c_9	c_{10}	c_{11}	c_{12}	$\geq z_{min}$

FIGURE 6-4.
The primal problem, showing constraint matrix.

assignment problem); all the other primal variables were nonbasic, or equal to 0, and this was indicated by default—that is, those x's for which no quota (or asterisk) was shown were assumed to be 0. Complete familiarity with the transportation algorithm of Section 3.2 will be required in Section 6.8, so you should perhaps review that now.

In Section 3.4 we also outlined roughly how a primal algorithm for the transportation problem would work, and you should probably also review that material now. We immediately achieved a feasible basic solution by making as-

u_1	u_2	u_3	v_1	v_2	v_3	v_4	
1	0	0	1	0	0	0	$\leq c_1$
1	0	0	0	1	0	0	$\leq c_2$
1	0	0	0	0	1	0	$\leq c_3$
1	0	0	0	0	0	1	$\leq c_4$
0	1	0	1	0	0	0	$\leq c_5$
0	1	0	0	1	0	0	$\leq c_6$
0	1	0	0	0	1	0	$\leq c_7$
0	1	0	0	0	0	1	$\leq c_8$
0	0	1	1	0	0	0	$\leq c_9$
0	0	1	0	1	0	0	$\leq c_{10}$
0	0	1	0	0	1	0	$\leq c_{11}$
0	0	1	0	0	0	1	$\leq c_{12}$
α_1	α_2	α_3	β_1	β_2	β_3	β_4	$\leq w_{max}$

FIGURE 6-5.
The dual problem, showing constraint matrix.

signments in such a way that all discrepancies were reduced to 0. However, such a solution would not usually be optimal, and we searched for improvements by looking for closed chains of possible reassignment. Those elements to which we assigned quotas became basic variables, and all other elements were nonbasic variables, and you should now realize that each such reassignment through a closed chain amounted to an iteration just like one in the simplex. The amount transmitted through such a chain was the maximum possible, subject to the constraint that one of the quotas went to 0; that is (just as in simplex), the maximum possible so that just one of the basic variables became nonbasic. Simultaneously, one of the elements which had not previously had an assignment received one; that is, just one of the nonbasic variables became basic. What we were not able to explain to you properly at that time was the necessity for drawing the chain in such a way that one and only one element in the chain would be nonbasic (this would be the incoming variable). All other elements in the chain must be basic. Furthermore, degeneracy often creates significant problems with such chains, as we shall see below. We now turn to a further explanation of degeneracy in transportation and assignment problems.

6.2 DEGENERACY

We show first that all transportation problems are degenerate; next we show that many transportation problems have additional degeneracies; and finally we show that all assignment problems are highly degenerate. In the following sections we shall indicate what difficulties this leads to and how they can be resolved.

In Chapter 0 we pointed out that it is permitted to add equations to one another. We therefore ask you to add the first three equations in Figure 6-4. The result is

$$1 \quad 1 \quad 1 \quad 1 \quad 1 \quad 1 \quad 1 \quad 1 \quad 1 \quad 1 \quad 1 \quad 1 = \alpha_1 + \alpha_2 + \alpha_3$$

or

$$\sum x_j = \sum \alpha_i \tag{6-1}$$

The first form is specialized for our particular 3×4 transportation problem; the second form is general if we add up all the row constraints (that is, all the constraints which apply to the rows of Figure 6-1). Now if you add up the remaining constraint equations (the column constraints), you will get

$$1 \quad 1 \quad 1 \quad 1 \quad 1 \quad 1 \quad 1 \quad 1 \quad 1 \quad 1 \quad 1 \quad 1 = \beta_1 + \beta_2 + \beta_3 + \beta_4$$

or

$$\sum x_j = \sum \beta_k \tag{6-2}$$

Any problem in which $\Sigma \alpha_i$ is not equal to $\Sigma \beta_k$ is an infeasible problem; we pointed out in Section 2.7 that, where such excess of origin or destination requirements occur, it was necessary to add dummies until equality did obtain, so that the problem could be solved. Hence

$$\sum \alpha_i = \sum \beta_k \qquad (6\text{-}3)$$

which is the same as (3-2), and it follows that equations (6-1) and (6-2) are identical.

Now this is a degenerate situation. When we first met degeneracy early in Chapter 1, we showed that it arose whenever one equation in a set could be derived from other equations in the set. Clearly we can derive any one of the seven equations in Figure 6-4 from the other six. For example, if we add up the first three equations and subtract the next three, we will get the seventh equation. Thus, any one of the seven equations is superfluous, and the whole equation set is degenerate.

The total number of equations, excluding any which are superfluous, is equal to the number of variables which will be basic in the optimum solution, in accordance with Dantzig's Theorem. This number of nonsuperfluous equations is at most $m + n - 1$, and so the maximum number of positive variables in the optimal feasible solution to a transportation problem will be $m + n - 1$, where m is the number of rows (origins), and n is the number of columns (destinations).

If you will look at Figure 3-8 on page 75, you will see that $m = 3$, $n = 3$, $m + n - 1 = 5$, and we do have five quotas greater than 0, that is, five basic variables. In Figure 3-9, $m = 3$, $n = 4$, $m + n - 1 = 6$, and on the right-hand side of this figure you will see an optimal feasible solution with six basic variables. However, on the left-hand side of the figure you will see an alternate-optimal feasible solution with only five basic variables. Such additional degeneracy is always possible (but not necessary) in transportation problems when there is additional redundancy in the constraint equations. The redundancy in this case arises because $\alpha_1 + \alpha_2 = \beta_1 + \beta_2$; that is, the amount available at the first two origins (17) is exactly equal to the amount required at the first two destinations (17), and of course the remaining requirements fall into a separate subset, $\alpha_3 = \beta_3 + \beta_4 = 8$.

In solving such a problem by the primal algorithm, degeneracy arises exactly as it arises in the simplex algorithm: at a point where we have to look for a smallest number, and we find a tie—that is, two (equally small) smallest numbers. For example, if we were solving the problem of Figure 6-6, our first assignment would be made to the element in (1,1)—which, in case you forgot, means the element in row 1, column 1. The amount assigned would be the smaller of $\beta_1 = 5$ and $\alpha_1 = 11$, so we would assign 5. This would leave a discrepancy of six in the first row, and we would assign this to the element (1,2), leaving a discrepancy of six in the second column. Now we will devote our attention to the element (2,2). The amount to be assigned would be the smaller of the row and column discrepancies, but both of these discrepancies are six. Thus we would have a degeneracy.

	5	12	2	6
11	23	11	25	5
6	17	8	3	46
8	25	35	3	10

(*a*)

	0	0	2	6
0	5/23	6/11	25	5
0	17	6/8	3	46
8	25	35	3	10

(*b*)

FIGURE 6-6.
Appearance of degeneracy. (*a*) A transportation problem; (*b*) assignment of quotas by the primal algorithm.

Still another way of looking at this would be to examine Figure 3-9*b* for alternate-optimal solutions. We use the same closed chain as before, starting at (1,2), running through (1,4), (3,4), (3,1), (2,1), (2,2), and back to (1,2). To obtain a new basic optimal feasible solution, we must transmit through this chain as much as possible. This will be the smallest of the quotas (1,4), (3,1), and (2,2), and these quotas are 5, 5, and 6. Again there is a tie. When we transmit five units through the chain, two of the quotas go simultaneously to 0. One of these elements becomes a nonbasic variable, and the quota of the other must be replaced by an epsilon (ϵ); recall that ϵ is a very small number, maybe a billionth, which is still a positive number distinct from zero.

The important point is that every transportation problem has an essential degeneracy so that the maximum number of basic variables in the optimal feasible solution will be $m + n - 1$. Problems which have exactly this many positive variables in the optimal feasible solution are said to be *nondegenerate*; only if there are additional degeneracies, so that the total number of positive variables is less than $m + n - 1$, is the problem called a degenerate transportation problem. And it is these "degenerate" problems, where we must add epsilons, that lead to difficulties in the primal algorithm.

There may be several such coincidences, and in the extreme case of the assignment problem each new row supplies a new coincidence, since each α_i and each β_k is equal to 1. Thus, an $n \times n$ assignment problem has n degeneracies, and the total number of positive variables in the final solution is just n. To solve such a problem by a primal algorithm would require $n - 1$ epsilons, which would create an intolerable mess. This is why we use a dual algorithm to solve the assignment problem.

An efficient primal algorithm for the assignment problem has been developed by the American mathematicians Michel L. Balinski and Ralph E. Gomory, which is not bothered by degeneracy; that is, it does not require epsilons. However, like the modi method (which is described in Section 6.7), it uses both primal and dual variables. Specifically, it makes a first feasible (that is, primal feasible) assignment by the northwest-corner rule, which results in assigning all elements along the diagonal. It then improves the assignment, always maintaining primal feasibility. At every step it computes the dual variables. It is not dual feasible until the optimal solution is reached, because there are negative cost numbers in the matrix. Thus, it is in some sense really a primal-dual algorithm rather than a primal algorithm. The possibility of many different algorithms, or modifications of algorithms, is further discussed in Section 6.9.

6.3 ALTERNATE OPTIMA

We hope you are not confused by the unfortunate convention of calling transportation problems nondegenerate if they have no degeneracies beyond that inherent in all transportation problems. What we have just proved is that the primal transportation problem is always degenerate (even though it is sometimes called nondegenerate). Now, what does that tell us about the dual problem? Clearly it tells us that the dual problem always has alternate-optimal solutions. You remember that degeneracy is associated with 0s in the B-vector and alternate optima with 0s in the C-vector; so degeneracy and alternate optima are dual properties.

We could have noted ahead of time that the dual problem always has alternate optima. Remember that u_i is the amount subtracted from the ith row, and v_j is the amount subtracted from the jth column. Now suppose we have solved the dual problem and have a set of optimal u's and v's; that is, we have found a way of subtracting certain amounts from each row and column (some of the amounts subtracted may have been negative—which is to say we may actually have added to some rows and columns) so that all assignments can be made at zero cost. Now suppose we add 1 to every row and subtract 1 from every column. Clearly this doesn't change anything; all the costs stay the same, and in particular all the 0s to which quotas have been assigned remain 0s. But it does give us a different set of u's and v's; it makes each u_i smaller by 1 and each v_j greater by 1. Thus we have a complete new set of dual variables, which is also an optimal set and which is different. Of course instead of adding and subtracting 1 we could add and subtract 2 or 17 or -169 so that there are an infinite number of alternate-optimal solutions to the dual problem.

Strangely, this infinite set of alternate-optimal solutions causes no trouble—as is clear from the fact that we had no trouble with it in developing the assignment algorithm in Chapter 2 or the transportation algorithm in Chapter 3. You have the choice, in step 8 of either algorithm, of adding to the covered rows and subtracting from the uncovered columns or adding to the covered columns and subtracting from the uncovered rows. If you do it one way you will end up with one set of dual variables, and if you do it the other way you'll end up with a different set. If this bothers you, then you can adopt a standard form in which u_1 is always equal to zero; that is, when you have completed the problem, and have an optimal set of dual variables, you look at u_1; if it happens to be equal to 3 (as it was in our sample transportation problem, Figure 3-8), you simply subtract 3 from every u and add 3 to every v. This is the method we shall use later in Section 6.7.

If the primal problem has additional degeneracies, the dual problem may have additional alternate optima. Recall that the additional degeneracy arises where there is a coincidence of subsets of origin requirements and destination requirements. If it happens that $\alpha_3 = \beta_1 + \beta_2$, then it may be possible to add only to the third row and subtract only from the first two columns to get still another alternate-optimal solution to the dual problem, for example, two different sets of optimal dual variables with $u_1 = 0$ in each set. An example was given in Problem 2-9 and is further explored in Problem 6-12.

Thus far we have been talking about degeneracy in the primal and the corresponding alternate optima in the dual; the same situation of course also works in reverse. If the primal problem happens to have alternate optima, then the dual will be degenerate. It happens that the primal problem of Figure 3-9 is degenerate and has alternate optima; it follows that the dual may have alternate optima and be degenerate. We say "may have" because a problem which has degenerate basic feasible solutions does not necessarily have degenerate optimal solutions. Note also that degeneracy of the dual of the transportation problem cannot be identified in any straightforward way with zero values of the dual variables, because of the infinity of alternate-optimal solutions. Only the $c_{ij} - u_i - v_j$ will show an extra zero (above the $m + n - 1$) if the primal has alternate optima.

6.4 THE INTEGER PROPERTY

It makes no sense in the assignment problem to assign a worker fractionally to a job. The nonlinear formulation (2-1) specifically excluded this (in the fifth line), but the linear formulation (2-2) does not. We blithely assumed that this wouldn't happen; that is, we assumed that if we solved the linear problem in which we omitted the integer constraints, we would not end up assigning $\frac{1}{3}$ of the seventh worker to the second job and the other $\frac{2}{3}$ to the ninth job. We asserted that this wouldn't happen because the optimum solutions always turn out to be integers, and we promised to prove that. We're going to prove it now, but first we wish to convince you that this was by no means a trivial matter. We do this by giving an example in which you can't get away with it; in this example, if you ignore the integer (nonlinear) constraints and solve the corresponding linear problem, you simply come up with the wrong answer.

6.4.1 The Subsystem Test Problem

In this problem we have a system consisting of several subsystems, each of which must be tested. For example, we may wish to test a hi-fi system; the various subsystems include the tuner, the amplifier, the speakers, and so forth. We also have several tests available: The first test may be to apply an overvoltage and see what happens, the second test might involve vibrations, and so forth. Each test costs a certain number of dollars to apply, and each test tests some of the subsystems but not others. For example, if we apply the overvoltage test and the system survives, then we may assume that the tuner and amplifier are both adequate and have been amply tested, but we will know nothing

Subsystem No. \ Test No.	1	2	3
1	1	1	0
2	1	0	1
3	0	1	1
Cost	1	1	1

FIGURE 6-7.
A 3 × 3 subsystem test problem.

about the speakers. *The subsystem test problem*, therefore, is to select a group of tests at minimal cost which tests each subsystem at least once. As we shall see, this is a mathematical programming problem, but it is not a linear programming problem because of the integer constraints.

We must first define a set of variables. These variables will be a set of x's, where each x_j will be 1 if the jth test is applied and 0 if it is not applied. The corresponding costs will be the c_j's, where c_1 is the cost of applying the first test, and so forth. The activity coefficients (the A-matrix) will consist exclusively of 1s and 0s; each $a_{ij} = 1$ if the ith subsystem is tested by the jth test and 0 if the jth test does not adequately test the ith subsystem. Our objective function will be to minimize $\Sigma c_j x_j$, that is, the total cost of applying all the chosen tests. There will be m constraints, one for each subsystem; each constraint will be of the form $\sum_j a_{ij} x_j \geq 1$, which is equivalent to saying that that subsystem must be tested at least once. In other words, somewhere in the row, an x_j has to be equal to 1 (that test has to be applied), where the a_{ij} is equal to 1 (that test tests that subsystem). Now this certainly looks like a standard linear programming problem, and we could go ahead and apply the dual simplex method to it and get the solution; the trouble is that we will get the wrong answer, because we will end up with noninteger values of the x's.

To illustrate this we give the very simple example shown in Figure 6-7, in which there are three tests and three subsystems. The first test tests subsystems 1 and 2, but does not test subsystem 3, and costs \$1, and similarly for the other two tests. Now obviously there are three alternate-optimal solutions to this problem, because any two of the three tests will suffice. That is, we can apply the first and second test ($x_1 = x_2 = 1; x_3 = 0$), or the first and third, or the second and third; in each case, each subsystem will have been tested either once or twice (once is enough, but twice is acceptable), and the total cost in each case will be $z = \$2$. No single test will suffice, and clearly therefore there is no solution cheaper than \$2. However, if you formulate this as a linear programming problem as indicated above, the minimum-cost solution will turn out to be $x_1 = x_2 = x_3 = \frac{1}{2}$; $z = 1\frac{1}{2}$. That is, the linear programming formulation will lead us to recommend applying each test half a time, thus testing each sys-

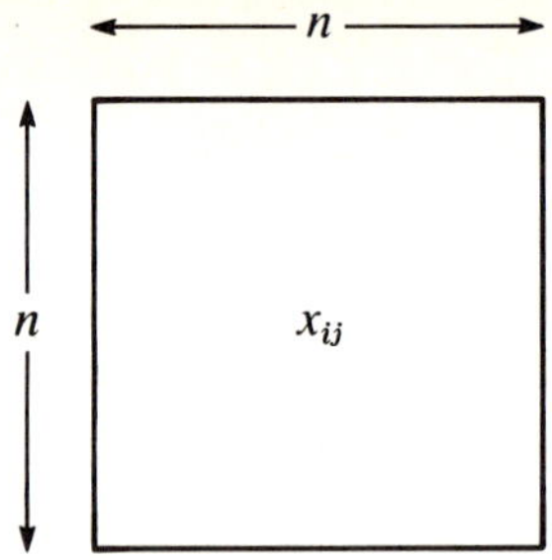

FIGURE 6-8.
Dimensions of primal-variables matrix for assignment problem.

tem exactly once at a total cost of $1.50. Mathematically this solution is perfectly satisfactory, but physically it is meaningless.

What we have shown is that integer constraints may not be lightly ignored. If the cheapest possible solution is a noninteger solution, the linear programming method will find that. The linear programming method works on real integer problems only if there is no noninteger solution which is cheaper than the best integer solution; in that case, when the linear programming method does find the cheapest solution, that solution will indeed be integer. Looked at another way, the linear programming method is too powerful; it finds the cheapest solution whether we want it to or not. It will therefore be useful to us only if the cheapest solution happens to be an integer solution.

In some problems, fractional solutions are acceptable. If the numbers are large, rounding off is reasonable—13,333 $\frac{1}{3}$ widgets or 13,333 widgets can hardly make any significant difference. And some variables admit of fractional values, for example, $x_1 = 3.2$ pounds or $x_5 = 2.7$ gallons. Note that the first sentence of Problem 4-11 asserts that noninteger solutions are permissible. But if the numbers are small and integers are required (subsystem test problem, Problem 4-16), then special techniques are required. One such technique is explained in Section 7.6. This technique is not especially difficult to learn, but solving large integer problems on a computer may require excessive time.

6.4.2 Proof of the Integer Property

We shall now prove that in the assignment problem the cheapest solution will always be an integer solution. The proof will proceed as follows. First we note that the solution consists of a set of x's. Second, we prove that in the optimal solution at most $2n - 1$ of these x's will be positive. Third, we prove that in the optimal solution there must be a column that has one x equal to unity and $n - 1$ x's equal to zero. And fourth, we prove that all other columns also have this *integer property*. This completes the proof, since it shows that no fractional x is possible in this optimal solution.

In Figure 6-8, we reproduce the primal-variables matrix (like Figure 6-1) for an $n \times n$ assignment problem, and in Figure 6-9 we reproduce the constraint matrix (like Figure 6-4) for the same assignment problem, indicating the dimen-

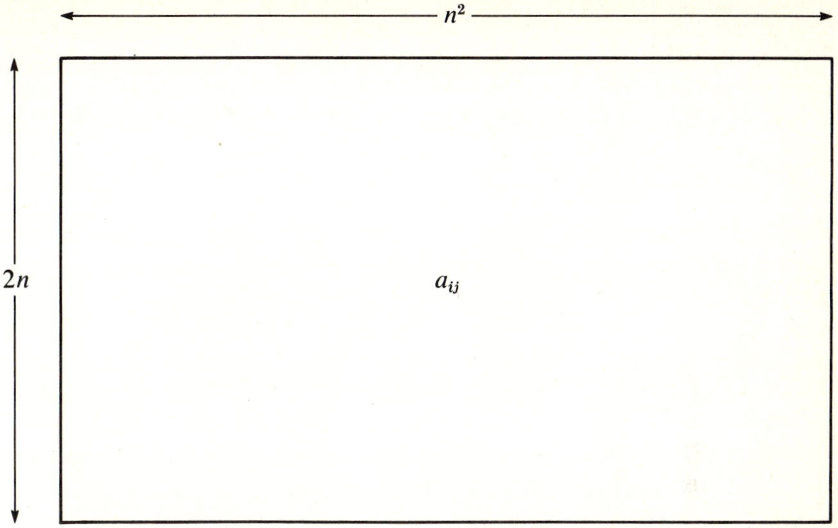

FIGURE 6-9.
Dimensions of constraint matrix for same assignment problem.

sions of these matrices. There are $2n$ equations represented in Figure 6-9, but one of these is superfluous, and so there are really only $2n - 1$ equations. It follows that in the optimal solution to this problem there will be at most $2n - 1$ positive variables; that is, when we have finished solving the problem, and thereby have found an optimal set of x_{ij}'s to put in the n^2 places in Figure 6-8, at most $2n - 1$ of them will be positive and all the rest will be 0.

Consider then this optimum set of x_{ij}'s in Figure 6-8, and try the assumption that each of the n columns in Figure 6-8 has at least two of these positive x's. This assumption is clearly impossible; there are n columns in Figure 6-8, and if each of these columns has at least two positive x's, there will be at least $2n$ positive x's; but we just agreed that in the optimal solution there are at most $2n - 1$ positive x's.

It follows that there must be at least one column in Figure 6-8 which has only one positive x. We redraw the matrix as Figure 6-10 and concentrate on this one column, whichever it is, that has only one positive x. The sum of all the numbers in the indicated column of Figure 6-10 must be 1 (you remember that every column of x's has to add up to 1), and since only one of these numbers is positive, that number must be 1; each of the other numbers in the column is 0. The 1 falls in some row, and whichever row it is we indicate it as in Figure 6-10. Now concentrate on that row that has the 1 in it. That row of x's also has to sum to 1, and so all the other elements in that row must be 0s. Now delete this row and column from the matrix, leaving an $(n - 1) \times (n - 1)$ matrix. This represents a new, smaller assignment problem to which we can apply exactly the same argument; that is, in this remaining matrix there must be at least one col-

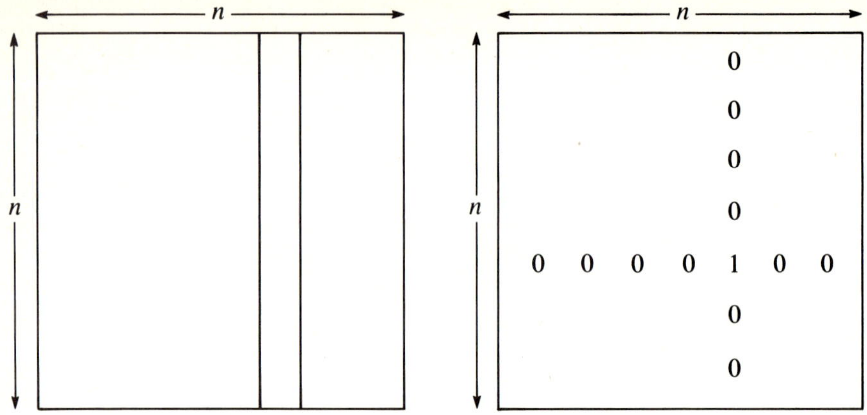

(a) (b)

FIGURE 6-10.
The integer property. (*a*) One column of the $n \times n$ matrix; (*b*) one column and one row of the matrix.

umn that has only one positive x. This x must be 1, and that 1 must be the only positive element in its row. We can then delete that row and column and continue. In that way we will show that in the optimal solution we end up with all 1s and 0s and no fractions, which is exactly what we wanted to prove.

Note the tremendous difference between the assignment problem and the subsystem test problem. In the latter the cheapest possible solution consisting of fractional x's is cheaper than the cheapest possible integer solution; the linear programming method finds that fractional solution, which is useless to us. In the former the cheapest solution (which we find with linear programming) is always an integer solution and is useful.

The simple but elegant proof which we have given above was discovered only recently. Because it illustrates how mathematics sometimes works, we thought you might be interested in the history of this problem. As you know, any matrix which satisfies the three constraints of (2-1) is called a permutation matrix. A matrix which satisfies the constraints of (2-2) also has a name. If it satisfies the last two constraints, it is called a *stochastic matrix* and is very important in probability theory, because the x's in each row can represent probabilities (which, as you know, are nonnegative and sum to unity). Any matrix which satisfies all three constraints of (2-2) is called a *doubly stochastic matrix*. An American mathematician named Garrett Birkhoff proved that the set of all $n \times n$ doubly stochastic matrices forms a convex polygon in n-dimensional space and that the vertices of this polygon are the $n \times n$ permutation matrices. When Dantzig subsequently proved that the solution of any linear programming problem lies on a vertex of the convex polygon of constraints, our theorem about the integer property of the assignment problem followed immediately.

6.4.3 Application to the Transportation Problem

Since the transportation problem is a special case of the assignment problem (any transportation problem which is not an assignment problem can be expanded, as in Figure 3-1, to an assignment problem with some identical rows and/or columns), we have also proved that the optimal quotas in any transportation problem will be integers if the origin and destination requirements are all integers. But we have actually proved a little more than this. Suppose all the origin and destination requirements were integral multiples of 100, that is, numbers like 300, 1700, and so forth; then the same proof shows that all the quotas in the optimal solution will be integral multiples of 100. You are asked in Problem 6-1 to prove this. This allows us a considerable short cut in solving such a problem, because we can simply throw away the hundreds and thus work with much smaller numbers and still be sure that we will get the right answer. In fact, we can add to our algorithm the following:

Step 0 Find the largest common factor of all α_i and β_j. Divide all α_i and β_j by this factor. Go to step 1.

We must then modify the final step so that we don't forget to multiply back by this factor when stating the final solution.

6.4.4 Application to the General Linear Programming Problem

You may be wondering whether there is some general rule which permits one to determine whether or not the solution to a linear programming problem must come out in integers. There is, but it requires some pretty complicated mathematics. We state the rule for you here, simply to show that it does exist.

Delete from the A-matrix rows, or columns, or both, or neither, in every possible combination of ways which leaves the same number of rows as columns. Compute the determinant of each resulting square matrix. If every such determinant is 1, 0, or -1, and if the original B-vector consists entirely of integers, then the optimal solution will come out in integers. [These conditions are sufficient but not necessary; that is, if the A-matrix and B-vector satisfy these conditions, then the optimal feasible solution will always come out in integers; but it is possible that the optimal feasible solution will come out in integers even if the matrix does not satisfy these conditions, as was true of our example problem (4-6) in Section 4.6.]

If you don't know what a determinant is, it doesn't matter very much; it's a number which can be computed for any square matrix in accordance with a certain formula. The point is that even if you did know how to compute a determinant, you couldn't actually do it in all possible ways according to the above rule. For even a fairly small transportation problem, say, 5×5, which leads to a 10×25 A-matrix, there are millions of ways of deleting rows and columns to lead to a square matrix. It's a bit like Koenig's Theorem, where there were an enormous number of ways of drawing lines covering rows and columns, but we didn't have to try them all in order to prove Koenig's Theorem. There are methods in matrix algebra for doing this kind of thing (Problem 6-21).

The real question is: If we start with the B-vector consisting entirely of integers, will it continue to consist entirely of integers as we go through simplex iterations? Suppose an A-matrix consists entirely of 1s and 0s, as it does in both the subsystem test problem and the transportation problem; what happens on the first iteration? Since our pivot cannot be a 0 (we consider only strictly positive a_{ij}'s in selecting the pivot in subroutine E), it will have to be a 1, and so our first operation will be dividing by 1 to get the new pivot row, that is, simply copying the row. The next set of operations will involve either copying a row (if a 0 is already where we need it) or multiplying the pivot row by 1 and subtracting it from various other rows. Thus, in developing the next tableau, all we do is subtract one row from another or copy a row, and in either case we will continue to have integers in the B-vector. However, this operation will introduce some -1s into the second tableau. Thus, in the second tableau, the A-matrix will consist exclusively of 1s, 0s, and -1s, and the B-vector will still be all integers. Now again in developing the new pivot row, we will choose a

1 and copy the row; however, in the next set of subtractions we might subtract a -1 from a $+1$ and develop a 2. Thus, in the third tableau, there might be a 2 somewhere in the A-matrix. If this turned out to be the pivot element, then we would have to divide that row by 2 in developing the fourth tableau, and this might lead to a fraction in the corresponding b. The point is that in the subsystem test problem this can happen; in the transportation problem it cannot happen. Whether or not it is going to happen depends upon the precise arrangement of 1s and 0s we start with, and this in turn can be determined by rules involving things like determinants.

In the transportation problem, if we solve it by means of the simplex algorithm, as we proceed we would never have anything in the A-matrix of successive tableaux except 1s, 0s, and -1s. Hence, the first operation in developing each tableau is simply copying the pivot row, and the succeeding operations consist either of copying the row if a 0 is already there (in which case the b—the value of the basic variable—doesn't change) or simply subtracting one row from (or adding one row to) another without any multiplications and divisions. If everything is as simple as this, the simplex algorithm is much too complicated a tool to use, which is why it is possible to develop special-purpose algorithms for the transportation problem which are much simpler than the ordinary simplex algorithm.

6.5 THE PRIMAL ALGORITHM

Now let us see how we would develop a simplified algorithm similar to the simplex for the transportation problem, working directly with the small matrix of Figure 6-1 and never bothering to write the larger matrix of Figure 6-4 at all. Recall that with the product-mix problem we started with the first basic feasible solution of making nothing, which was far from optimal; then we performed successive iterations, each of which (in the absence of degeneracy) improved things. Before each iteration we examined the c_j's and picked the most negative (though that was not essential—we could have picked any negative c_j). Then in the iteration, just one of the nonbasic variables became basic, the others remaining zero, and just one of the basic variables became nonbasic, the others remaining positive. The B-vector was the only thing we were really interested in, since it constituted the values of the basic variables, but we had to compute all the a_{ij}'s as a necessary evil.

In Section 7.3 we shall show that one doesn't really have to compute all the a_{ij}'s. For the transportation problem we don't have to compute any of them (they are all 0, 1, or -1 anyhow). And we don't have to compute the values of all the basic variables, because some of them don't change in any given iteration.

In Figure 3-10 and Section 3.4 we established a first feasible basic solution (which might have been far from optimal) by the northwest-corner rule. Then we examined the element (3,1) and found that we could increase its quota, changing other quotas appropriately through a closed chain, in such a way as to decrease cost by two units of money for each unit shipped. This two was the opportunity cost associated with this element; that is, the c_j for this nonbasic variable was -2. We then drew a chain which included the elements (1,1), (1,3), and (3,3) whose quotas would be changed. The other two basic elements and the other three nonbasic elements did not change their values, and no computations had to be made on them. The only computations required were the additions and subtractions in going through the chain.

Now you can begin to see why we had to be vague about the primal algorithm in Section 3.4. For each nonbasic variable we have to compute a c_j, which is done by adding and subtracting through a chain. And if this variable is then chosen as the incoming variable, all the other variables in the chain will change (that is, the x's will either increase or decrease), and so they must all be basic variables. And when we have degeneracy, this gets messy. In fact, in Figure 3-10 we had degeneracy, and one of the elements in our chain did not have a quota. So formally this element should have been given an ϵ quota. In the numerical example below we shall show exactly how this is done. The algorithm depends on the fact that if we have a nondegenerate basic solution (which may include ϵ's), then for each nonbasic variable there exists a unique closed chain in which all the other variables are basic. Our algorithm then consists of finding this unique chain for each nonbasic variable, computing the c_j, picking the most negative, then reestablishing the chain and performing the iteration. If it turns out that there are no negative c_j's (that is, if it turns out that for every nonbasic variable we cannot profitably transmit through its unique chain), then an optimal solution has been reached.

In carrying out this algorithm, we never change the original costs, which we call c_{ij}. At each iteration we compute, for each nonbasic variable, the opportunity cost, which we call c_j, by adding to the c_{ij} at odd-numbered elements of the chain and subtracting from the c_{ij} at even-numbered elements. We ask you to bear in mind this distinction between our use of the symbols c_j and c_{ij}.

As indicated in Section 7.3, we could avoid this ambiguity by denoting what we have here called c_j by $c_{ij} - z_{ij}$, but we have preferred to introduce as few symbols as possible.

This completes our informal presentation of the primal algorithm (we shall formalize it below). You should note carefully that what we are really doing is a simplex iteration, which turns out to be much simplified for two reasons. First, there are no multiplications or divisions—all operations on the primal variables (the quotas) are simply additions and subtractions. Second, as before, only one nonbasic variable needs to be changed, but this time most of the basic variables also don't change, and so it pays us to locate the chain which includes all of the elements which do change. Each of these differences is associated with the integer property of the transportation problem.

We will now illustrate the difficulties caused by degeneracy by solving the problem of Figure 6-6. We have already indicated how the degeneracy arises. Since we must have $m + n - 1$ basic variables in order for the algorithm to work (in order that there always exists an appropriate unique chain), we must add an epsilon to either the row or column discrepancy. In this case we choose to add it to the row discrepancy, and thus put the epsilon quota on the element

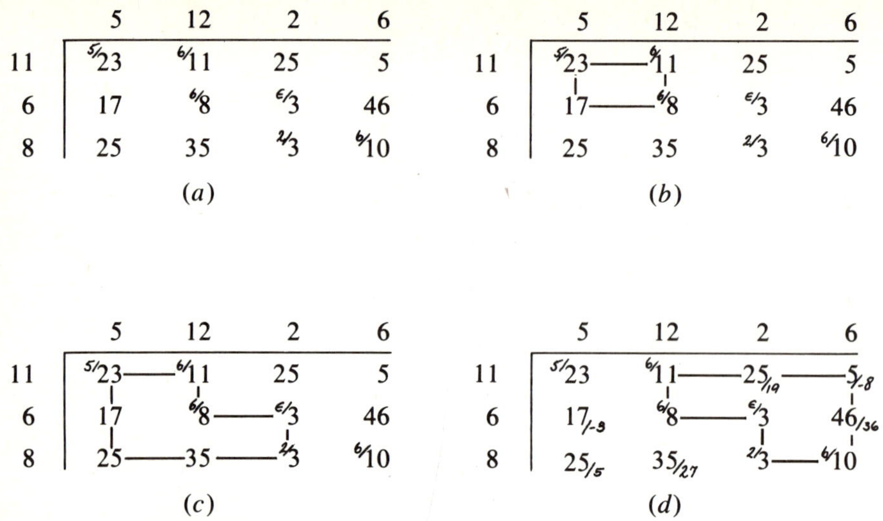

FIGURE 6-11.
Solving by the primal algorithm. (*a*) Completing the first feasible assignment; (*b*) the unique closed chain for (2,1); (*c*) the unique closed chain for (3,1); (*d*) completion of c_j calculations.

in (2,3), which we have done in Figure 6-11*a*, although we might equally well have put it on the element in (3,2). We then complete the assignment, as shown in Figure 6-11*a*. Note that the quota of (3,3) is really $2 - \epsilon$, but we write 2, because 2 and $2 - \epsilon$ are not distinguishable.

We now pick a nonbasic variable, and in accordance with our rule we will start with the topmost one in the leftmost column. Since the element in (1,1) is basic, the topmost available nonbasic element is (2,1). We develop its unique closed chain as shown in Figure 6-11*b*. Please convince yourself that this chain is indeed unique; that is, there is no other closed chain involving the element in (2,1) with all other elements being basic (positive quotas). To determine the c_j for this element, we assume that we add 1 unit to its quota; that is, we change its quota from 0 to 1, which will cost us 17. We must now subtract one unit from the quota of the element in (1,1), which will save us 23. Continuing around the chain, we see that we must add a cost of 11 and subtract a cost of 8. The total net change in cost in going through this chain will be $17 - 23 + 11 - 8 = -3$, and we enter this c_j in Figure 6-11*d*; we place it below and to the right of the cost in order to distinguish it from the quotas, which we have placed above and to the left.

We next examine the nonbasic variable in (3,1), and this time the chain is a little more complicated, as shown in Figure 6-11*c*. Incidentally, this chain includes our epsilon; note that if we had chosen to put the epsilon in the alternate position, we would have gotten a different chain and therefore a different c_j. For this particular chain, to transmit one unit through the chain will cost us $25 - 23 + 11 - 8 + 3 - 3 = 5$, and we enter the 5 as shown on Figure 6-11*d*.

You can see that this is a tedious business. You have to repeat this for each of the nonbasic variables, and there could be quite a lot of them. For example, for a 10×20 transportation problem, there would be 200 primal variables, of which 29 would be basic and 171 would be nonbasic. We would have to establish the chain and compute the c_j for each of these 171 nonbasic variables. Some of these chains may be very complicated (like the convoluted chain in Figure 2-10). In this case we have simply shown you the final result in Figure 6-11*d*, but naturally we urge you to compute each of these numbers for yourself to make sure you understand how to find the chain and how to go through it.

Now we will iterate on the element with the most negative c_j, namely, (1,4). The question arises: How much can we transmit through this chain? The most we can transmit is the smallest of the quotas for the even-numbered elements of the chain—namely, 6, ϵ, and 6—and so the most we can transmit is ϵ. This is frustrating; it means that all we do is move the epsilon from its present location in (2,3) to (1,4), because, for example, $6 - \epsilon = 6$. Now since at least some of the chains will have changed, we have to go through the whole mess again and recompute all the c_j's. Do you begin to understand why degeneracy can cause trouble in the primal transportation algorithm?

Of course there is an alternative. The very first negative c_j we found, namely, the -3 in (2,1), could have been used immediately. We can transmit up to five units through this chain, saving three units of money for each unit transmitted, for a total improvement of the primal objective function by 15. However, we'll probably sooner or later have epsilon troubles in any case. We ask you in Problem 6-2 to complete the solution of this problem by the primal algorithm.

We are now prepared to present the formal

Algorithm for the Transportation Problem (Primal)

Step 0 Find the largest common factor of all α_i and β_j. Divide each α_i and β_j by this factor. Choose the element (1,1), that is the element in the northwest corner. Go to step 1.

Step 1 Examine the row discrepancy and column discrepancy of this element; if they are the same, go to step 2; if they are different, go to step 4.

Step 2 Assign the amount of these discrepancies as the quota of this element. If this is the element in the southeast corner—that is, the element (m,n)—go to step 5; otherwise go to step 3.

Step 3 Assign the quota epsilon to the element in the next column, same row; now go to the element in the next row, same column as the element to which you have assigned the epsilon, and go to step 1.

Step 4 Assign the smaller of these discrepancies as the quota of this element, thus reducing this discrepancy to 0 and leaving the other discrepancy positive. If the positive discrepancy is the row discrepancy, choose the element in the same row, next column, and go to step 1; if the positive discrepancy is the column discrepancy, choose the element in the next row, same column, and go to step 1.

Step 5 If every element has either a quota or a c_j, go to step 7; otherwise, choose an element which has neither, and go to step 6.

Step 6 Locate the unique closed chain starting at this element, going vertically to an element with a positive quota, horizontally to an element with a positive quota, and so on, returning eventually to the element where you started. Now compute c_j as follows: Add the cost of this starting element, subtract the cost of the next element in the chain, add the cost of the next element in the chain, and so on until the chain has been completed. Enter this c_j; go to step 5.

Step 7 Choose the element with the smallest (most negative) c_j. If this c_j is negative, go to step 8; if it is 0 go to step 9; if it is positive, go to step 10.

Step 8 Reestablish the chain for this element. Find the smallest of the quotas of the even-numbered elements of the chain; note whether there is a tie for smallest element. Now go through this chain, adding this smallest quota to the quotas of the odd-numbered elements of the chain (that is, starting with the non-basic variable) and subtracting it from the quotas of the even-numbered elements. Now note whether there is more than one zero quota in the chain (there will be at least one 0, and there will be more if and only if you noted a tie above). If so, change all but one of these 0s to epsilons. Check to be sure that all discrepancies are zero. Erase all c_j's and go to step 5.

Step 9 The solution is optimal, but there may exist alternate-optimal solutions. If you are not interested in these, go to step 10. If you are interested, record the present solution for processing in step 10, and then examine each c_j whose value is 0. Reestablish the chain for this element; if there is no epsilon quota for an even-numbered element of the chain, there exists an alternate-optimal solution, which you should develop by applying step 8. Now go to step 10.

Step 10 The solution is complete. Erase all epsilons. Transfer the quotas to the original problem, remembering to multiply back by any factor by which you may have divided in step 0. Check the solution. Stop.

In step 3, you may prefer to put the epsilon below instead of to the right. In either case, you will return to step 1 with the element below and to the right.

In step 5, you have your choice of going to step 7 (and 8) immediately when you find a negative c_j, or as soon as you find a particularly attractive (large) negative c_j, or not until you have computed all c_j's.

Note that you do not actually change the α's and β's in this problem (if you want, you may mark the discrepancies lightly in pencil while going through steps 2 and 4, then erase the pencil marks when you first get to step 5). Thus, they are available for checking each time you complete step 8, to make sure you didn't make an error. Checking simply means seeing that the sum of quotas in each row adds up to its α, and similarly for columns.

We have not stated what you must do if you find an epsilon as the smallest quota on an even-numbered element of the chain. The answer is that you do what step 8 says, which then simply results in moving the epsilon; that is, adding an epsilon to (or subtracting it from) a quota different from 0 does not change that quota.

It is essential that you keep things lined up so that you do not confuse the quotas (the x_{ij}'s), the original costs (the c_{ij}'s), and the opportunity costs (the c_j's). The best way to do this is to write the costs in ink, and then put the quotas and c_j's (which you are going to change from time to time anyhow) in pencil. When the quota goes to 0, do not write the 0, but simply erase all evidence of a quota at that point; the idea is that when you are ready to go to step 7 every cost element will have either a quota or a c_j so that you will be able to tell quickly by examining the matrix whether to go to step 6 or step 7. If you indicate a zero quota, you may forget to compute the c_j for that element.

6.6 SHADOW PRICES

We have asserted that the dual variables of any linear programming problem are the shadow prices; that is, the dual variable associated with a particular row is equal to the value to us (within the constraints of this particular problem) of an extra unit of the resource which constrains that row. But for the transportation problem we have shown that the optimal dual variables are arbitrary, in the sense that you may select any one dual variable and assign it any value you like (we have suggested that you select $u_1 = 0$). Obviously shadow prices cannot be arbitrary; that is, the value of a resource to us cannot depend on what value we happen to assign to u_1. However, if we decide to change the optimal dual variables, we must select some number, add it to every u_i, and subtract it

1	7	4
−4	2	−1
1	7	4

FIGURE 6-12.
Shadow prices for the transportation problem of Figures 3-3 and 3-8.

from every v_j so that any particular sum of a u_i and v_j is invariant; thus, if we add 3 to every u and subtract 3 from every v, a particular sum such as $u_5 + v_2$ will not change. It is these sums which are the shadow prices of the transportation problem.

Look again at the final solution to our first example transportation problem on Figure 3-8. In this solution, $u_1 = 3$, $v_1 = -2$, and the sum $u_1 + v_1 = 1$. Now we could change u_1 to 0 if we wanted to, but if so, we must change v_1 to 1, and $u_1 + v_1 = 1$ is still true. This is the shadow price associated with the element (1,1) of the problem; it is the price of increasing the supply at the first origin and the requirement at the first destination, each by one unit. In Figure 6-12 we compute this sum $u_i + v_j$ for every element of the problem from Figure 3-8. The cost of increasing α_1 and β_1 by one each would be 1 dollar (or other unit of money); but if we increase the supply at the second origin by one unit, and increase the requirement at the first destination by one unit, the new problem would actually be 4 dollars cheaper than the old, because the shadow price is -4.

Now recall the apparent paradox of Problem 3-2. We asked the question whether increasing the requirements could actually give us a cheaper solution and found to our surprise that it was indeed possible. What we should have asked was: Are negative shadow prices possible in a transportation problem? In most simple linear programming problems, negative shadow prices are not possible, because the shadow prices are the dual variables, and all variables are constrained to be nonnegative; however, you remember that the nonnegativity constraint on a dual variable goes with an inequality constraint in the primal which is in the "right direction." If the inequality constraint in the primal goes in the wrong direction, the dual variable is actually constrained to be nonpositive (row 9 of Figure 5-1). And if the constraint in the primal is an equality, the dual variable is unconstrained; that is, it does not have to be nonnegative—it may come out positive, zero, or negative (row 10 of Figure 5-1). In the case of the transportation problem, all constraints in the primal are equalities, and so all dual variables are unconstrained; hence it should be no surprise if we end up with negative dual variables in the optimal solution.

Another way of looking at this is an alternative formulation of the primal transportation problem, in which all origin constraints are in the form of less-than-or-equal-to inequalities, and all destination constraints are in the form of greater-than-or-equal-to inequalities. This says that you can ship no more from any origin than is available, and you must ship to each destination at least as much as is required there. The solution, of course, comes out exactly the same, provided the origin and destination requirements are balanced. Since the transportation problem involves minimization of cost, each of the destination constraints goes "the right way," and so each of the dual column variables (the v_j) is a nonnegative variable; but each of the origin constraints goes "the wrong way," and so each of the dual row variables (the u_i) is constrained to be nonpositive. From this point of view, our original question might have been: In the optimal solution, is it possible that the most negative of the row dual variables added to the least positive of the column dual variables yields a negative sum? Put this way, the result hardly appears to be a paradox.

The terms "opportunity cost" and "shadow price" are not specialized terms in linear programming but are widespread in economics. Each is a special case of a "marginal cost," which is a fundamentally important economic concept. An opportunity cost is the change in objective function caused by a unit change of a decision variable. In terms of a product-mix- or diet-type problem, and using the notation of calculus, it is dz/dx_j (or, more precisely, $\partial z/\partial x_j$). A shadow price is the change in objective function caused by a unit change in a constraint (resource or requirement); that is, dz/db_i (or, more precisely, $\partial z/\partial b_i$).

It is important to note that these are both systems concepts; they tell the overall effect of such a change on the system. In other words, if you change an x_j or b_i and then reoptimize, you will obtain this new value of the objective function.

Now in connection with a product-mix- or diet-type problem, we pointed out that if the x_j is a slack variable, the opportunity cost is equal to the shadow price. This is a theorem arising from the theory of duality. In the transportation problem, this is no longer the case. We have given one heuristic reason above (any particular dual variable is arbitrary); let us give two more: The transportation problem doesn't have any slack variables; and the derivative $\partial z/\partial \alpha_i$ is not well defined, because if you change an α_i you must change some β_j. When you specify which β_j, you can then define the shadow price as $u_i + v_j$. The opportunity cost is, of course, $\partial z/\partial x_{ij}$ as before.

6.7 THE MODI METHOD

The primal algorithm for the transportation problem would be very neat if it weren't for the messiness of having to compute the c_j's for all the nonbasic variables through the awkward chains. It turns out that it is possible to compute the c_j's without the chains, using the help of the dual variables. The method of accomplishing this is called the *modi* (for *mo*dified *di*stribution) *method*.

We must review what we did in the dual algorithm (of Chapter 3) and the primal algorithm (of Section 6.5), and why we did these things, in terms of what we know about linear programming and the simplex method. In the dual algorithm we worked primarily with the dual variables, where u_i was the amount subtracted from the ith row and v_j was the amount subtracted from the jth column. At each iteration (each time we went through step 8) we increased the (weighted) sum of dual variables. The solution was never feasible until we got to the end; that is, only at this point were the sums of quotas equal to the requirements and the constraints of the original problem thereby satisfied. However, it was always "optimal" because we never assigned a quota except at zero relative cost. Thus we used the primal variables (the quotas) to help us

(a)

					u_i
	5/23	6/11	25	5	0
	17	6/8	ε/3	46	
	25	35	2/3	6/10	
v_j	23				

(b)

					u_i
	5/23	6/11	25	5	0
	17	6/8	ε/3	46	−3
	25	35	2/3	6/10	−3
v_j	23	11	6	13	

FIGURE 6-13.
Solving by the modi method. (*a*) Assigning the dual variables; (*b*) complete set of dual variables.

determine how we should manipulate the dual variables. What we are going to do now is analogous: We are going to use the dual variables to help us determine how to manipulate the primal variables. However, since this will still be a primal algorithm, we will start exactly as before to obtain a feasible assignment of quotas (that is, a feasible set of primal variables). Thus, the first four steps of the algorithm are identical to those in Section 6.5.

Hence we start, on Figure 6-13, with the point which we had reached at Figure 6-11*a*, which is the point we shall have achieved when we first get to step 5. Consider the element in (1,1). It has a cost $c_{11} = 23$ and a quota $x_{11} = 5$. It is a basic variable (because it has a positive quota), and so its relative cost must be $c_1 = 0$, and so $u_1 + v_1$ must be equal to 23. Why? Because $c_{ij} - u_i - v_j$ is a relative cost (as we demonstrated early in Chapter 2 when we first subtracted from rows and columns), and this relative cost is precisely what we have been calling c_j. Hence if $c_{ij} = 23$ and $c_j = 0$, we must have $u_i + v_j = 23$. Now recall that any one of the dual variables is arbitrary, and we have chosen to set $u_1 = 0$ as our arbitrary choice. If this is true, v_1 must be equal to 23, and we have entered $u_1 = 0$ and $v_1 = 23$ in Figure 6-13*a*.

For exactly the same reason, v_2 must be 11, because $c_{12} - u_1 - v_2 = 0$, or $11 - 0 - 11 = 0$; we have therefore entered $v_2 = 11$ on Figure 6-13*b*. In a basic feasible solution (which may include epsilons, as this one does) there are always exactly $m + n - 1$ basic variables (that is, elements with positive quotas), and each one of these permits us to determine the value of precisely one new dual variable. Continuing on Figure 6-13*b*, we note that the element in (2,2) has a cost of 8. Since this is a basic variable, we must get to zero when we subtract u_2 and v_2 from 8. We already know that $v_2 = 11$; it follows that $u_2 = -3$, and we enter the -3 on Figure 6-13*b*. Continuing in the same fashion, the element in (2,3) is a basic variable (because it has a quota, even though that quota is an epsilon), which together with $u_2 = -3$ permits us to determine $v_3 = 6$. Continuing in this way we complete the tabulation of the u's and v's in Figure 6-13*b*.

Now suppose we went ahead and subtracted each of the u's from its appropriate row and each of the v's from its appropriate column; we know that the cost of each basic element would go to 0; what would happen to the costs of the nonbasic elements? Well, for example, the element in (2,1) would go to -3, and the element in (3,1) would go to 5. And these numbers are precisely the c_j which we seek, obtained without the use of chains. If you continue this computation of $c_{ij} - u_i - v_j$ for each nonbasic element, you can fill in all the c_j's and find that the result is identical with Figure 6-11*d*.

We have asserted that this solution is feasible (because all the quotas by rows and columns add up to the corresponding requirements) but that it is not optimal; this is from the primal point of view. From the dual point of view we can say that the solution is not feasible, because when we subtract the u's and v's that we now have, some of the cost elements go negative. When we finally get to the point where none of the c_j's is negative, then we will have achieved dual feasibility and the problem will have been solved.

Algorithm for the Modi Method

Step 0 through step 4 Identical to primal algorithm in Section 6.5.

Step 5 Erase all dual variables except u_1, which equals 0; go to step 6.

Step 6 If all the dual variables have been determined, go to step $\frac{1}{2}$; if not, choose an i and a j such that x_{ij} is positive (that is, such that this element has a quota—which might be ϵ), and either u_i is determined and v_j is not, or v_j is determined and u_i is not. Determine this dual variable by using the equation $c_{ij} - u_i - v_j = 0$. Enter this u_i (or v_j). Go to step 6.

Step $6\frac{1}{2}$ If every element has either a c_j or a quota, go to step 7. If not, choose an element which has neither; determine $c_j = c_{ij} - u_i - v_j$; enter this c_j; go to step $6\frac{1}{2}$.

Step 7 through step 10 Identical to primal algorithm above (except of course that the first word of step 8 must now be "establish" instead of "reestablish," and at the end of step 8 you go back to step 5 of the modi algorithm instead of step 5 of the original primal algorithm).

As indicated in step 11 on page 77, you should perform all possible checks: (1) $\sum_j x_{ij} = \alpha_i$ and $\sum_i x_{ij} = \beta_j$; that is, row by row and column by column, the quotas add to the original discrepancies. (2) $x_{ij}(c_{ij} - u_i - v_j) = 0$; that is, for every element, either the quota or the relative cost is zero (or in degenerate cases, both); this is called the *principle of complementary slackness* and is discussed in Section 7.7. (3) $\Sigma\alpha_i u_i + \Sigma\beta_j v_j = \Sigma\Sigma c_{ij}x_{ij}$; that is $w_{\max} = z_{\min}$. (4) $x_{ij} \geq 0$ and $c_{ij} \geq u_i + v_j$; that is, both primal and dual solutions are feasible (1 above is also needed for primal feasibility).

We apologize for "step $6\frac{1}{2}$," but we wanted to keep the numbering the same as before. It is worth noting that the numbering of steps in an algorithm (or instructions in any problem-oriented computer language) is arbitrary.

The modi method is much easier to carry out than the primal algorithm given above, either by hand or by computer. It is quite difficult to write computer-program instructions for establishing the chains. What we've learned once again is that the theoretical insights which we have gained permit us to develop a simpler algorithm; and, once again, these theroretical insights are related to the theory of duality applied to linear programming problems.

6.8 SENSITIVITY ANALYSIS OR POSTOPTIMALITY ANALYSIS

Postoptimality analysis implies that after we have located the optimal solution, we wish to do still further analysis. This additional analysis may be prompted by the fact that something has changed or by the fact that we were uncertain about one of the numbers in the first place and wish to find out how a change in that number would affect the solution. This latter view of the problem is called *sensitivity analysis*; that is, we wish to find out how sensitive our solution is to change in some of the input data.

The things that may have changed are the origin and destination requirements (α_i and β_j) and the costs (c_{ij}). Or we may have opened a new warehouse, which represents an entirely new origin (that is, a new row, complete with its appropriate α and all the costs) or a new destination (new column). We assume that if the origin and/or destination requirements change, they are still in balance—equation (6-3)—or we know how to supply the appropriate dummies. Whatever the change, we will make the corresponding change in our final matrix and hope to be able to reoptimize with very little work even in a large problem—perhaps just a single application of step 8 of the primal algorithm of Section 6.5 or step 7 of the dual algorithm of Section 3.2. You should be sure you are thoroughly familiar with all these algorithms before you proceed.

6.8.1 Change of Requirements

We have a complete set of optimal primal variables and a complete set of optimal dual variables; this will be true if we have solved either by the dual algorithm of Chapter 3 or by the modi method of Section 6.7. Suppose, to be specific, that we add 1 to the requirements at the second origin and first destination of our original transportation example (Figures 3-3 and 3-8); that is, we now have $\alpha_2 = 4$ and $\beta_1 = 2$. We already know from Figure 6-12 that we will be able to find a new solution of this larger problem which is \$4 cheaper than the old one; but what is this new optimal solution? Simply go back to Figure 3-8, which we have copied in the upper left corner of Figure 6-14, and solve this new problem. You may solve it by either the primal or the dual algorithm; but whichever one you choose, you should note that the starting point of Figure 6-14 represents an almost-solved problem.

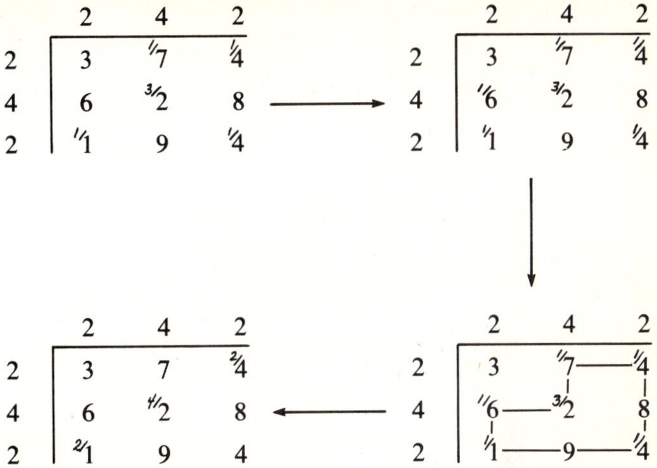

FIGURE 6-14.
Postoptimality primal algorithm.

If you choose to solve it by the primal algorithm, you will immediately increase the quota of the element in (2,1) from 0 to 1; this adds 6 to the cost. But now there exists a closed chain consisting entirely of elements with positive quotas, through which we can transmit that one-unit quota, saving $6 - 2 + 7 - 4 + 4 - 1 = 10$. Thus we have expended 6 and saved 10, or saved a net total of 4. You might note that this new solution is highly degenerate.

If you wish to solve it by the dual algorithm, it is more convenient to use the optimal values of $c_{ij} - u_i - v_j$ which were shown in the last matrix of Figure 3-7 and which we have copied as the first matrix of Figure 6-15. We now cover the columns with discrepancies of 0 and prime the uncovered 0 (second matrix). Checking the discrepancy of this row we find that it is 0; so we cover the row, star the twice-covered essential 0, uncover its column, and prime the new uncovered 0 (third matrix). Once more we find that the discrepancy in the row is 0, and so we cover its row, star the twice-covered essential 0, and uncover its column. Now we prime the uncovered 0 in (2,2), we find that the discrepancy is positive (last matrix in Figure 6-15), we have a complete chain, and we get the same reassignment (and therefore of course the same new solution) as before.

We have shown how the modified solution can be obtained if there is a change in origin or destination requirements: We may get the new answer by applying either the dual or the primal algorithm to the almost-solved problem which results by making this minor change. We will thus learn the complete new set of primal variables and dual variables as well as the new value of the objective function. If we are interested in learning only the change in the objective function, then it is even easier; we have only to look at the shadow price.

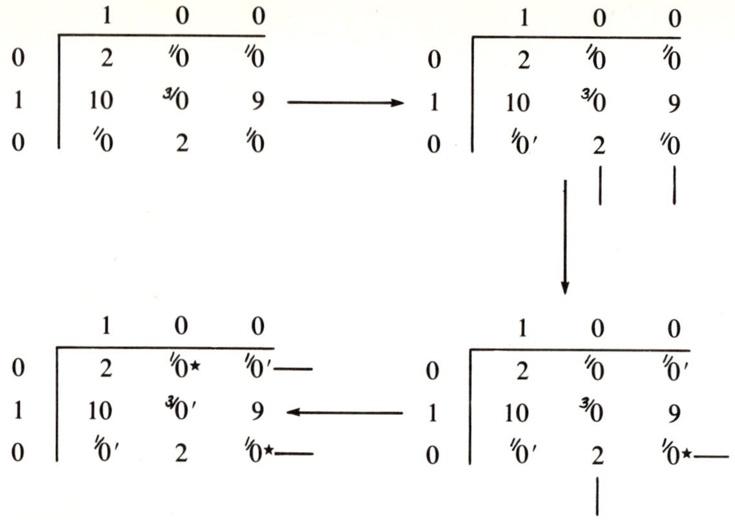

FIGURE 6-15.
Postoptimality dual algorithm.

6.8.2 Change of Cost (Nonbasic Variable)

Similar conclusions apply if one of the costs, c_{ij}, is changed. In this case there is a considerable difference whether the c_{ij} corresponds to a basic or a nonbasic variable. Consider first the simpler case of a nonbasic variable.

Looking once more at our original transportation problem (Figure 3-3, 3-8, or 6-14), suppose the cost in (1,1) were changed from 3 to some other number. How would this affect the solution of the problem? At the present time this element has a quota of 0; that is, it is sufficiently expensive that our optimum solution does not involve making any shipment via this route. Clearly, if the cost were even higher, we would certainly not use this route. Thus, the first conclusion of our sensitivity analysis is that if $c_{11} > 3$ our solution remains unchanged.

Now suppose the cost is less; at some point it would pay us to use this cheaper route, and just what this point is becomes immediately apparent by looking at the first matrix of Figure 6-15. If c_{11} had been 2 in the original problem instead of 3, then c_{11} would be 1 on Figure 6-15 instead of 2. This is a most important point to note; all the operations on the c_{ij} in reaching Figure 6-15 have been simple additions and subtractions. So if the original c_{11} were reduced by 1, then this final c_{11} would be reduced by 1. And if this number were still greater than 0, then we would still not make any assignment to it. An extension of the same line of thinking indicates that if the original cost were reduced from 3 to 1, this cost would be 0 in Figure 6-15, and we might have an alternate-optimal solution. In this particular case there would be such an alter-

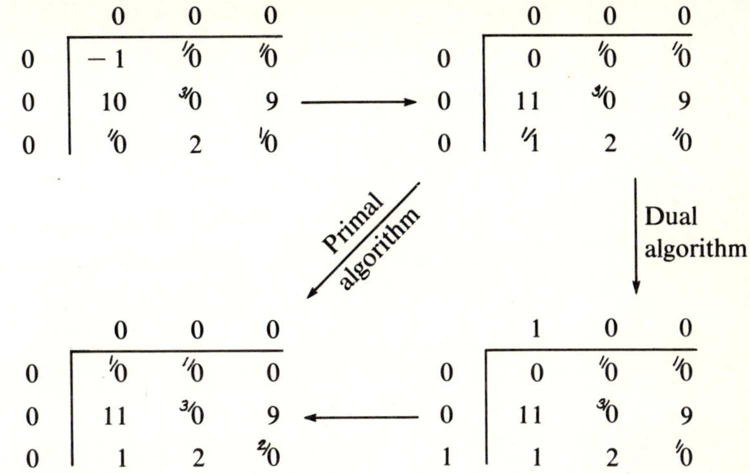

FIGURE 6-16.
Sensitivity analysis—nonbasic variable.

nate-optimal solution, since there exists a closed chain involving this element and elements with positive quotas (which will always be true if the solution is nondegenerate).

Finally, if the original cost had been $c_{11} = 0$, then in Figure 6-15 we would have $c_{11} = -1$, which we have shown in the first matrix of Figure 6-16. This is an infeasible solution, and we must now add 1 to either the first row or the first column. Suppose we choose to add 1 to the first column (second matrix). This is again infeasible because we have a quota assigned to a nonzero cost; or if you prefer to look at it from the primal viewpoint, it is nonoptimal. You may prefer to consider this an almost-solved problem by the primal algorithm, in which case you will immediately note that the c_j for the element (1,1) is -1, leading to the reassignment (and final solution) shown in the lower left corner of Figure 6-16. Alternatively, as we have done in the lower right, you can remove the quota from the positive cost element, creating discrepancies, and then proceed to solve by the dual algorithm. If so, you will cover the two columns with zero discrepancies, prime the zero in (1,1), star the two other zeros in the first row, uncover both columns, prime the zero in (3,3), and find a chain of reassignment leading to the final solution shown in the lower left. Again in this case there's an important conclusion about sensitivity which is easily determined by inspection—that is, by comparing the original c_{ij} matrix (Figure 6-14) with the matrix of $c_{ij} - u_i - v_j$ in the optimal solution (Figure 6-15). You can then immediately see that the present solution is optimal as long as $c_{11} \geq 1$. If $c_{11} = 1$, we will have a new alternate-optimal solution; if $c_{11} < 1$, the present solution is no longer optimal. If you now wish to ask the further question (exactly what does happen if $c_{11} < 1$?), then you must make the appropriate changes and apply either the primal or dual algorithm to the resulting almost-solved problem.

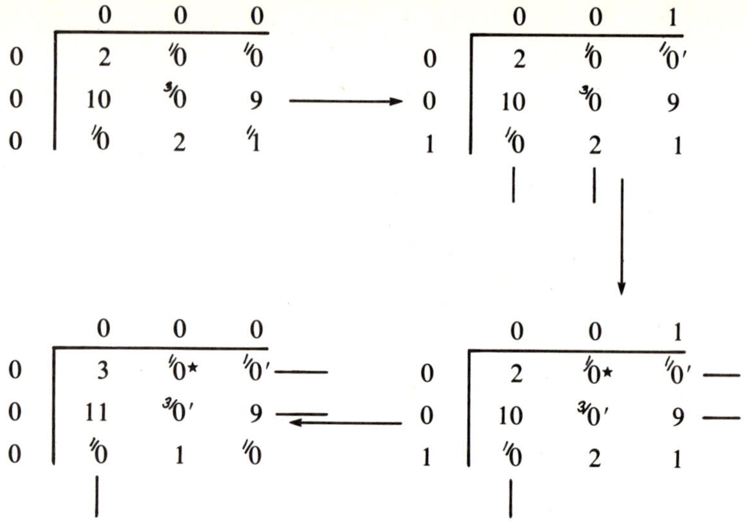

FIGURE 6-17.
Sensitivity analysis—basic variable (increasing cost, dual algorithm).

6.8.3 Change of Cost (Basic Variable)

Now suppose we change the cost element $c_{33} = 4$. In the final solution, the corresponding variable is basic; that is, $x_{33} = 1$, which means that we are using this route. Clearly if the cost of this route became sufficiently high we would avoid using it. What is not so obvious is that if the cost of this route became much lower, that might also alter our optimal solution, because in that case we might take advantage of that lower cost by increasing the quota of this cheaper route to a number larger than 1.

Our basic analysis is the same as before. Suppose first that we increase c_{33} from 4 to 5; the corresponding change in the $c_{ij} - u_i - v_j$ matrix would be to increase c_{33} from 0 to 1 as indicated on Figure 6-17. Now solving by the dual algorithm we remove the quota from (3,3), increasing the corresponding discrepancies, cover the two columns with discrepancies of 0, and prime the uncovered 0 (second matrix in Figure 6-17). We cover this row, star the twice-covered essential 0, uncover the second column, prime the 0 in (2,2), cover this row, and find that all our 0s are covered (third matrix). The smallest uncovered element is now the 1 in (3,3), and the adding and subtracting leaves the lower left matrix, where we have made the appropriate assignment. Now compare this last matrix carefully with the first. You will see that the solution to the problem has not changed. The cost has increased by 1 because our quota of 1 in (3,3) is now applied to a higher cost. Note that the element in (3,2) is now 1 instead of 2 and that all the other numbers in the cost matrix are either unchanged or have increased. It is clear that if we repeated this same

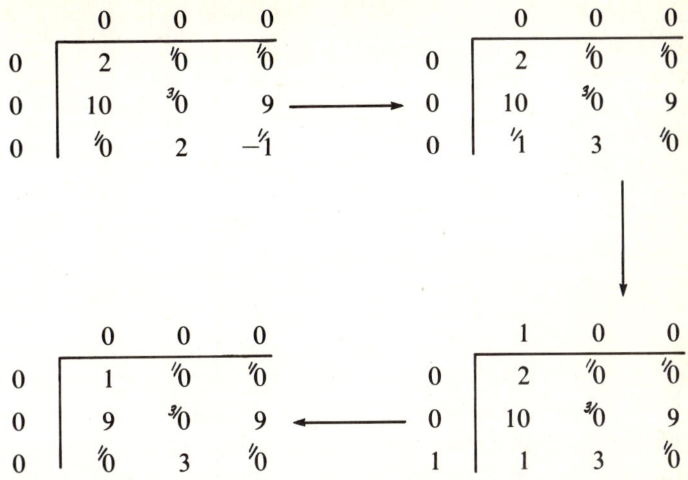

FIGURE 6-18.
Sensitivity analysis—basic variable (decreasing cost, dual algorithm).

procedure, we would again make the same changes in the matrix; that is, if the original c_{33} were now increased from 5 to 6, the new matrix corresponding to the lower left of Figure 6-17 would have all the numbers either the same or increasing, except for the element in (3,2), which would decrease one more, namely, to 0. At this point we would have a new alternate-optimal solution. Thus we may say that the present solution remains optimal as c_{33} increases from 4 to 6; if c_{33} increased beyond 6 we would get a new solution in which x_{33} would be 0, x_{32} would be 1, and there would be some other corresponding changes.

Now let's see what happens if we decrease c_{33}. Suppose we reduce it from 4 to 3. The analysis by the dual algorithm, which should now be reasonably familiar, is shown in Figure 6-18. The cost c_{33} has become -1, which is infeasible (in the dual sense), so we add 1 to the third row, yielding the second matrix. We must now remove the quota from this element, increasing discrepancies, at which point we note that we may immediately subtract from the first column and reassign the quotas as in the last matrix. Comparing the two matrices on the left of Figure 6-18, we again see that the optimal solution has not changed (of course the total cost has been reduced by 1), and all cost elements have either remained the same or increased, except this time for the elements in (1,1) and (2,1). Again a further application of the same procedure—that is, reducing c_{33} from 3 to 2—results in the same changes, and at this point the element in (1,1) would go to 0, yielding an alternate-optimal solution. If c_{33} were reduced below 2 we would have a new optimal solution, in which the quota x_{33} would be increased to 2, the quota x_{11} would be increased from 0 to 1, and other quotas would be correspondingly decreased.

Of course you can get the same result from the primal algorithm. In Fig-

3_{1}	${}^{1}7$	${}^{1}4$	0
6_{9}	${}^{3}2$	8_{9}	-5
${}^{1}1$	9_{3}	${}^{1}3$	-1
2	7	4	

(*a*)

3_{2}	${}^{1}7$	${}^{1}4$	0
6_{10}	${}^{3}2$	8_{9}	-5
${}^{1}1$	9_{2}	${}^{1}4$	0
1	7	4	

(*b*)

3_{3}	${}^{1}7$	${}^{1}4$	0
6_{11}	${}^{3}2$	8_{9}	-5
${}^{1}1$	9_{1}	${}^{1}5$	1
0	7	4	

(*c*)

FIGURE 6-19.
Sensitivity analysis—basic variable (primal algorithm). (*a*) Decreasing cost; (*b*) original cost; (*c*) increasing cost.

ure 6-19 we show the calculation of all c_j by the modi method: in the center for the original problem, on the right for the modified problem where c_{33} is increased, and on the left for the modified problem where c_{33} is decreased. On the right, all opportunity costs are the same or bigger except for the one in the third row, second column. Clearly if c_{33} increases to 6 (or beyond) this opportunity cost will go to zero (or negative), and the new solution will be found by reassignment through the closed chain (3,2), (3,3), (1,3), (1,2). On the left, all opportunity costs are the same or bigger, except for those in the first and second rows of the first column. If c_{33} decreases to 2 (or beyond) the opportunity cost in (1,1) will go to zero (or negative), and the new solution will be found by reassignment through the closed chain (1,1), (3,1), (3,3), (1,3). These conclusions are, of course, the same as those we reached through the dual-algorithm analysis. You have your choice of which approach you like; in some problems one will be easier, and in others the other will be easier.

Our sensitivity analysis has thus shown that the original solution (of Figure 3-8) is optimum, provided $2 \leq c_{33} \leq 6$. If $c_{33} = 2$ or 6 there would be alternate optima; outside this range there would be different optimal solutions. Of course when we say the solution remains optimum, we mean the optimal values of the primal variables remain unchanged. The total cost changes by the product of the quota (in this case $x_{33} = 1$) and the change in the c_{ij} (in this case c_{33}).

We have shown several things: If there should be a change in any of the numbers that describe the problem, then there may or may not be a change in the solution. Depending upon the exact change that is made, the answer may be easy to determine or somewhat more difficult. In many cases, part of the answer can be determined by inspection; for example, by inspection of Figure 3-8 we can determine the range of c_{11} over which the solution is optimal. To determine exactly what the new solution will be requires additional labor, but it is important to note that we do not have to start the problem all the way from the beginning! The change results in an almost-solved problem which is comparatively easy. This ease is not unique to little toy problems like the one we have been playing with. In a large practical problem, say, with 100 origins and several hundred destinations, the solution process would obviously require the services of a computer—preferably a large one. But the postoptimality analysis for any single change could easily be done by hand. Of course if we wanted the effects of changes in many of the constants of the problem, it would be desirable to program the computer to do this for you. We ask you in Problem 6-22 to write such an algorithm.

All these conclusions (with appropriate modifications) are also valid for the general linear programming program: Given the final tableau, achieved by either the primal or dual algorithm, a change in any cost or requirement results in an almost-solved problem. Some questions can be answered by inspection; others require an iteration, in which one algorithm or the other may be easier to work with, depending on the change. Again, a cost change for a basic variable is a little more complicated than for a nonbasic variable. But in any case, for the general linear programming problem, the sensitivity analysis is a little more complicated, and we defer it to the next chapter.

6.9 FANCY ALGORITHMS

The primal algorithm of Section 6.5 consists of three parts: steps 1 through 4, which obtain a first feasible solution; steps 5 through 8, which improve the solution; and steps 9 and 10, which process the optimal solution. The particular method used in Section 6.5 for the first part is called the *northwest-corner rule* and for the second part the *stepping-stone method*. The modi method is a modification of the stepping-stone part of the algorithm. We have also indicated that you can modify steps 5 through 7 of the algorithm (either stepping-stone or modi) by picking any nonbasic variable which offers a cost improvement rather than searching through all of them.

Modification is also possible in steps 1 through 4, in which we pick a first feasible solution. The simplest modification is to choose low-cost elements rather than starting in the northwest corner. This will usually (but not always) result in getting a better first solution, thereby requiring less iterations of steps 5 through 8. We have not written out this algorithm formally, because we think the idea is pretty obvious. When you do have all the discrepancies reduced to 0 by this method, you'll have to count the number of basic variables. If it exceeds $m + n - 1$, you've made a mistake somewhere, which can be easily rectified by finding an appropriate closed chain of basic variables through which you can transmit so that one of them goes to 0. If the number of basic variables is less than $m + n - 1$, you must add epsilons. You will probably put them in the wrong place, which means that for some nonbasic variable(s) the chain will not exist, while for others it may not be unique. You will then have to move one or more epsilons. (There is a rule for how to place them to avoid this, if you want to go to the trouble to learn it.) Other than that, the algorithm is the same.

A more complicated but more efficient method of choosing a first feasible set of quotas is by looking at penalties. A penalty represents what it costs you not to assign any quota to a particular inexpensive route and is measured by the difference between the cheapest cost in a particular column and the second cheapest cost (or cheapest and second cheapest cost in a particular row, or both). For example, in our first transportation problem, Figure 3-3, if we do not use the 2 in the second column, then the best we can do is to use the 7 in that column, which involves a penalty of 5. According to this method, therefore, the best place to start with the first feasible assignment of quotas would be to assign as much as possible to this element, namely, 3; and you will remember that in the final solution, 3 was in fact assigned to this element. The concept of penalties can be formalized and leads to certain types of "branch-and-bound algorithms" which are discussed in Section 7.6.1.

Whether such improvements in the algorithm are worthwhile depends on the costs involved. If you plan to run the problem on a computer, the extra cost in programming the extra steps may not be worth the small amount of computer time saved by iterating fewer times in later steps in the algorithm. This depends upon the sophistication of the techniques and the number of times the problem will be run. If you plan to run the program many times per day in controlling an operating system, even a small saving in computer time is worthwhile.

One obvious improvement that you might wish to make in certain problems is to divide all the costs by a constant. For example, dividing all the costs by 100 is equivalent to changing the unit of measurement from cents to dollars, which is clearly permissible. If all the costs were given in cents, and yet all were integral numbers of dollars, then dividing by 100 (that is, knocking out the last two zeros), would save a good deal of work. Any particular problem which you have may possess a special structure which suggests a modification of the algorithm. For example, suppose all the costs in the matrix were between 100 and 200; subtracting 100 from every element is certainly permissible (this is how we started in the dual algorithm) and in such a case would reduce all the cost numbers from three digits to two, which would surely make life easier.

The point is that lots of algorithms are possible, and you are free to develop your own modifications, provided they are valid. The validity of a proposed modification should be easy to determine if you understand the theoretical aspects of the problem.

6.10 IMPORTANT TERMS

The following important terms have been defined in this chapter. You should not go on to Chapter 7 until you thoroughly understand each of them. Each is in italics where it is first defined.

- Subsystem test problem
- Integer property (of assignment and transportation problems)
- Modi method
- Postoptimality analysis
- Sensitivity analysis

Problems

6-1 Show that if all the α_i and β_j in a transportation problem have a common factor, then all the quotas in every basic optimal feasible solution of that problem will have the same factor.

6-2 Complete the solution of the problem of Figure 6-11 by the primal algorithm of Section 6.5. Develop all alternate-optimal solutions if any.

6-3 Consider

	5	3	4	6	8
5	8	6	3	14	24
12	0	7	6	0	14
9	30	6	9	14	6

(*a*) Solve by the modi method.
(*b*) Check (by showing $z_{\min} = w_{\max}$).
(*c*) List a complete set of optimal values for the primal variables.
(*d*) Are there alternate optima? If so, list one set.
(*e*) List a complete set of optimal values for the dual variables.
(*f*) Are there alternate optima? If so, list one set.
This problem is continued in Problems 6-4 and 6-11.

6-4 In Problem 6-3, what happens if

(*a*) c_{11} is changed from 8 to something else?

(*b*) c_{12} is changed from 6 to something else?

(*c*) α_1 and β_1 both increase from 5 to 6?

(*d*) β_2 is increased from 3 to 4 and α_2 is increased from 12 to 13?

Each part requires multiple answers: In parts (*a*) and (*b*), over what range does the solution not change, and how does it change if you go outside that range; in parts (*c*) and (*d*), what is the change in cost, and how is that new cost achieved?

(*e*) If you did part (*a*) correctly, you found an alternate optimum when $c_{11} = -1$ and a new solution when $c_{11} < -1$. However, as c_{11} continues to decrease, this new solution changes. Explore this.

(*f*) Similarly, what happens in part (*b*) if c_{12} continues to decrease?

Note: The real costs are 100 more than shown; for example, $c_{11} = 108$. We have subtracted 100 from each cost to make the problem easier. This explains why you must consider negative values for the original costs—because $c_{11} = 99$ now comes out $c_{11} = -1$.

6-5 In the first paragraph of Section 6.4.1 we described the "subsystem test problem" and gave as an example testing a hi-fi system. In the second paragraph we developed a mathematical model of this problem and illustrated it in Figure 6-7. We then showed that this model could not be solved by linear programming techniques but required integer programming techniques, such as those described in Section 7.6. Now we ask a different question.

(*a*) Suppose you had a good integer programming algorithm; would you believe the answer? That is, do you think the mathematical model developed in the second paragraph and displayed in the figure is an adequate model of the problem described in the first paragraph?

(*b*) If not, can you think of a problem for which this would be a good model?

Drill Problems

6-6 Suppose in Problem 3-6 that each supplier had specified an intention to reject any contract if the award were less than half the amount of the bid. What do you do now? How is the total cost affected? *Hint*: This is formally very difficult, as indicated in Section 7.6, but actually rather easy.

6-7 Solve by any algorithm you like:

	9	5	3	3	9	5	2	2
1	8	4	3	7	9	0	6	1
8	3	6	6	7	1	0	0	8
7	0	7	2	8	5	9	0	7
4	1	0	1	5	8	3	8	7
7	5	5	1	9	6	8	9	7
2	5	3	8	1	2	9	1	3
9	5	1	8	6	3	2	6	8

Answer: Minimum cost = 58.

6-8 Consider

	30	50	20	40	30	10
50	2	1	3	3	2	5
40	4	2	2	4	3	4
60	6	5	4	2	4	1
30	4	3	3	1	2	2

(*a*) Solve by the dual algorithm.

(*b*) Now that you know the answer, it is very easy to solve the original problem by the primal algorithm of Section 6.5. Do it.

(c) Is the problem degenerate (beyond the usual degeneracy of all transportation problems)? How do you know? How else do you know? (One answer from the problem, one from the solution.)

6-9 (a) Solve by the modi method:

	6	12	9	5
16	15	23	76	19
8	41	8	16	23
8	29	16	28	31

(b) Give the complete set of optimal primal variables.
(c) Compute $z_{\min}$ from this set.
(d) Does there exist an alternate-optimal basic primal solution? If yes, give one; if no, how do you know?
(e) Give the complete set of optimal dual variables.
(f) Compute $w_{\max}$ from this set.
(g) Does there exist an alternate-optimal set? If yes, give one; if no, how do you know?
(h) Is the optimal primal solution degenerate? How do you know?
(i) Does there exist a feasible nonoptimal primal solution which is degenerate? If yes, give one; if no, how do you know?
(j) Does there exist an optimal nonbasic primal solution? If yes, give one; if no, how do you know?
(k) Does there exist a feasible nonoptimal nonbasic primal solution? If yes, give one; if no, how do you know?

6-10 Continuing Problem 6-9,
(a) What is the new value of $z_{\min} = w_{\max}$ if $\alpha_1 = 17$ and $\beta_1 = 7$, other numbers as in Problem 6-9?
(b) How is this new value achieved?
(c) What is the new value of $z_{\min} = w_{\max}$ if $\alpha_2 = 9$ and $\beta_1 = 7$, other numbers as in Problem 6-9?
(d) How is this new value achieved?
(e) What happens if c_{22} changes from 8 to other nonnegative numbers? (All α's and β's as in Problem 6-9.)
(f) Suppose all numbers were really 100 bigger than shown, so c_{22} could continue to decrease below zero.

Extra Drill Problems

You can use any of the problems in Chapter 3 as extra drill problems, or you can make up your own drill problems. See the note after Problem 2-6, and make sure $\Sigma\alpha_i = \Sigma\beta_j$ (or add dummies). You can also practice postoptimality analysis on any of these problems. Pick a parameter (c, or an α and β), change it, explore what happens, then check by showing that your solution is correct for some specific value. For example, suppose you pick c_{23} which equals 8, and find a new optimal solution at $c_{23} = 4$. Show (by the modi method) that for this problem with $c_{23} = 4$ this is indeed the correct solution and that there is an alternate-optimal solution (which you have found).

Advanced Problems

6-11 In each part of this problem, you are given a complete problem (all c_{ij}, α_i, β_j) which is balanced ($\Sigma\alpha_i = \Sigma\beta_j$), nondegenerate, and has no alternate optima.
(a) You are given a set of optimal primal variables; how do you find a set of optimal dual variables?

(*b*) You are given a set of optimal dual variables; how do you find a set of optimal primal variables?

(*c*) In the optimal solution, $x_{23} = 0$; however, the boss insists that you send three units via this route anyhow for public-relations reasons. How do you find

(i) The extra cost of this decision?

(ii) The new optimal allocation with this constraint? (Assume $\alpha_2 > 3$ and $\beta_3 > 3$.)

(*d*) After you have written out your answer to (*c*), which is tricky, try it on Problem 6-3.

6-12 (*a*) In Problems 2-4 and 2-9 we found the same unique optimal primal solution but had different final matrices. Show (via sets of alternate-optimal dual variables) how to go from one final matrix to the other by adding to and subtracting from rows and columns.

(*b*) In exploring alternate optima in Section 2.4 and Problem 2-12, we assumed that further additions and subtractions were not relevant. Explore this.

6-13 (*a*) Write a computer program for solving transportation problems. Debug it on one of our small problems, then try it on the following matrix (the minimum cost is 58). We think the dual algorithm will be easier to program, but do as you like. See the hint in Problem 2-8.

	4	4	6	4	3	2	7	3	8	6	4	2
2	2	3	9	2	3	5	8	7	0	2	2	2
8	9	3	3	7	5	5	3	9	7	7	3	2
3	8	7	2	5	2	1	2	8	0	6	2	4
7	9	0	3	2	6	4	9	7	6	7	6	3
9	3	2	0	4	0	5	6	9	3	0	1	6
6	6	3	7	6	3	5	8	7	0	3	0	4
5	7	0	9	6	9	2	5	2	0	6	7	9
4	9	3	6	1	2	8	5	2	7	0	0	5
9	5	2	7	7	6	5	1	5	3	3	5	9

(*b*) Generate some rules for picking a better first solution. Program them. Run the new program. How are time and storage requirements affected? [Merrill Flood has described several such rules for the dual algorithm in *Naval Research Logistics Quarterly*, vol. 8 (1961), no. 2 (Sept.), pp. 257-276.]

6-14 (*a*) Continuing Problem 3-8, suppose that you are not allowed to store more than one substance in a particular facility (because, for example, each substance requires a different controlled environment). Show that the problem will then often be insoluble.

(*b*) Now suppose that the above restriction applies to the first n facilities, but there is an $(n + 1)$st facility which (at higher cost) can store as much of anything as you like. Show that this problem is always soluble.

(*c*) Show that these problems are not transportation problems and cannot be solved by any of the algorithms we have developed. Write an exact mathematical description of these problems [similar, for example, to equations (3-1)].

This problem is continued in Problem 7.5-2.

6-15 The three-dimensional assignment problem is as follows: Given a three-dimensional array of numbers c_{ijk}, select a set of elements whose sum is minimum under the constraints that appropriate sums must be equal to 1. Specifically, for example, we are supposed to assign workers ($i = 1, \ldots, n$) to crews ($j = 1, \ldots, n$) to aircraft ($k = 1, \ldots, n$): We are given a set of n^3 costs, each representing the cost incurred if the ith worker is assigned to the jth crew and this crew in turn is assigned to the kth aircraft. Or each c_{ijk} might represent the cost of locating the ith warehouse in the jth city in the kth year if we can only build one warehouse per year. Other applications are given in the *Journal of the Canadian Operations Research Society*, July, 1967, p. 72, and in *Operations Research*, March, 1968, p. 423.

(*a*) Formulate mathematically in a form equivalent to (2-1).

(*b*) Show that the solutions will not in general come out as integers. *Note*: The way to show this is to generate a "counterexample" in which the minimum-cost integer solution will have higher cost than the minimum-cost solution. This can be done with n as small as 2 and each c either 0 or 1. See if you can develop a pair of 2×2 matrices (which represent the c_{ij1} and c_{ij2}, respectively) which have this property.

6-16 The three-dimensional transportation problem. A firm produces p different items ($k = 1, \ldots, p$) in m different factories ($i = 1, \ldots, m$) which must be shipped to n different warehouses ($j = 1, \ldots, n$).

(*a*) The jth warehouse requires exactly A_{jk} items of type k; the ith factory produces exactly B_{ik} items of type k; and the route from factory i to warehouse j must carry exactly C_{ij} items in total [this last restriction is obviously unrealistic, but we will modify it in part (*b*)]. It costs a total of c_{ijk} to manufacture an item of type k in the ith factory and ship it to the jth warehouse. Formulate as a three-dimensional transportation problem; that is, define a set of variables x_{ijk}, and then write formulas corresponding to (3-1) and (3-2).

(*b*) To make the problem more reasonable, the jth warehouse requires at least A_{jk} items, the ith factory can produce at most B_{ik}, and the route can carry at most C_{ij}. Reformulate. You will find that the constraints corresponding to (3-1) are easy to formulate as the appropriate inequalities, but some of the ones corresponding to (3-2) are not so easy. To see this, convert the constraints corresponding to (3-1) into standard form by adding slack variables, then add these same slack variables to convert the relations between A's, B's, and C's into equations. You will see that some of these equations lead to obvious inequalities, while others do not.

6-17 (*a*) Write the mathematical description of both the primal and dual forms of the transportation problem [corresponding to (3-1) and (3-3)] for the formulation discussed in the last paragraph of Section 6.6. Note that, corresponding to (3-1), there will only be inequalities; and in (3-3), all variables are constrained.

(*b*) Investigate the duality of degeneracy and alternate optima in this formulation. In particular, consider the problem:

	5	5	2
4	1	2	3
4	4	5	6
4	7	8	9

Show that the primal has 18 alternate-optimal basic feasible solutions but no basic feasible degenerate solutions. Show that the dual has an infinite number of optimal feasible solutions: In one of these, one of the u_i or v_j is zero; in none is more than one of the u_i or v_j zero; and in none is any $u_i + v_j$ zero.

(*c*) There is a paragraph early in Section 7.7 on *complementary slackness.* Read that paragraph, then discuss how it applies (or how it is modified) for the present formulation.

6-18 Consider the following optimal solution to Problem 3-5 (optimal x_{ij}'s):

1	0	0	6	1
0	4	5	0	0
2	0	0	0	6

Note that the number of positive variables is $m + n - 1$. Is this solution basic or nonbasic?

6-19 Consider a typical transportation problem—say Problem 3-5. Suppose you added random three-digit numbers to each cost and kept the requirements the same; or suppose you kept the costs the same and added random three-digit numbers to most of the requirements (and adjusted the others); or suppose you changed costs and requirements to three-digit numbers. Would these problems be harder? Much harder? You should consider carefully what one means by "hard." And you should try it.

6-20 (*a*) The set of primal variables at the end of step 7 is always ________________ and ________________ but never

____________________ and/or ____________________ until the last time through step 7. The set of dual variables selected in step 6 of the modi algorithm is never ____________________ until the end. In the algorithm of Chapter 3, the solution was always ____________________ and ____________________ but never ____________________ and/or ____________________ until the end. Fill in each blank with either primal-feasible, primal-optimal, dual-feasible, or dual-optimal. Primal-feasible means satisfying all the constraints of (3-1); dual-feasible means satisfying all the constraints of (3-3). What do primal-optimal and dual-optimal mean?

(*b*) What principle always holds throughout both algorithms?

6-21 (If you know determinants.)

(*a*) The 3×3 matrix of Figure 6-6 has nine 2×2 and nine 1×1 submatrices. Show that just one of these 19 matrices has a determinant different from 1, 0, or -1.

(*b*) Show that each square submatrix of the 7×12 matrix of Figure 6-4 (or 6-5) has a determinant of 1, 0, or -1. *Hint*: Consider any 7×7 square submatrix. Show that it has at least one row or column with all 0s or with only one 1. Expand in cofactors along that row or column. Then its determinant is 0, or 1 or -1 times the determinant of some 6×6 submatrix.

6-22 Write out in algorithmic form the rules we derived in Section 6.8.3 for change in cost of a basic variable. It is assumed that the original problem, with m rows and n columns, has been solved and that you have optimal sets of x's and of u's and v's (with $u_1 = 0$). You are to explore the results of changing a particular c_{ij} over the range from very high (so that $x_{ij} = 0$) to very low (so that x_{ij} is as large as possible); that is, as in Problem 6-4*f*, you should explore beyond the first change in x_{ij}. *Hint*: It is probably easiest to use the method indicated in Figure 6-19. You may refer to some of the steps of the algorithms in Sections 6.5 and 6.7 as subroutines.

7
SOME SPECIALIZED AND ADVANCED TOPICS

If you have mastered the preceding chapters of this book, you have learned quite a lot about the assignment and transportation problems—in fact most of what is important to know about these problems—although there are a number of other models, interpretations, and algorithms, some of which are presented in Section 7.4. These two problems are special cases of the linear programming problem, and you have learned enough about linear programming to construct models of most such problems, to solve many such problems, and to interpret the results of the computer solution of most such problems. There remains one class of problems which require a special trick for their solution. We present this trick in Section 7.1, and when you have learned that you will be able (in principle) to solve any linear programming problem, although of course large ones cannot really be solved without the aid of a computer. In Section 7.2 we extend our concepts of postoptimality analysis to the general linear programming problem, but as you will see there are no new principles involved. In Section 7.3 we present another algorithm for solving linear programming problems, which is more complicated in the sense that it is more difficult to understand, but simpler in the sense that it may take less time to run. For the latter reason it is the one often used in the solution of real problems, and it might be desirable for you to know about it in case you have to work with computer output on such problems.

Linear programming itself is a special case of a larger class of techniques called *mathematical programming,* and in the last few sections of this chapter we briefly describe certain problems which cannot be handled by our linear methods and indicate some algorithms for solving such problems. Finally, mathematical programming itself is a subset of the quantitative techniques used in the field called operations research, and at the end of the book we discuss these briefly.

The sections in this chapter may be read in any order, though perhaps the order given is best. Appropriate problems will be found at the end of most sections and (in Section 7.4) subsections.

7.1 ARTIFICIAL VARIABLES

In a typical maximization problem, such as the product-mix problem, all the constraints are of the less-than-or-equal-to type; such a problem can generally be thrown directly into a feasible canonical form for immediate solution by the simplex algorithm. Similarly, in a typical minimization problem, such as the diet problem, all the constraints are of the greater-than-or-equal-to type; such a problem can generally be thrown directly into an "optimal" canonical form for immediate solution by the dual simplex algorithm. However in certain cases, namely, those in which the constraints include inequalities in both directions, or equations, there is no immediate method of going directly to either a feasible or an "optimal" canonical form.

The theoretical difficulties of such a problem mirror the practical difficulties. Suppose we have a large product-mix problem with numerous products and numerous resource constraints of the less-than-or-equal-to type, but that we also have some constraints of the greater-than-or-equal-to type representing contractual commitments to supply certain amounts of materials to certain customers. We may not be able to determine easily whether the problem is insoluble or soluble; that is, the constraints may be so constraining (Figure 4-19*c*) that there is no feasible solution, or (as in Figure 4-19*b*) there may be a region of feasible solutions with many vertices, and we have to determine the optimal one. The algorithmic difficulties can be seen in terms of the bedbug analogy of Section 4.8. As long as the bedbug is on the surface of the chandelier, he can always find an edge to go along which will remain on the surface of the chandelier and which will improve his situation; if he can't he's at the optimum. But now assume that all the edges are extended in the form of wires surrounding the chandelier, with many vertices out in the infeasible region. Suppose he finds himself on one of these infeasible vertices; remember he's nearsighted. He might wander around among these infeasible vertices indefinitely and never find his way back to the chandelier.

Certain methods have been developed for handling these difficulties. They work with a simplex-like tableau and go back and forth between simplex iterations (using a column with negative c_j) and dual simplex iterations (using a row with negative b_i). However, they are neither as easy nor as generally useful as the methods described below.

To get out of this difficulty, we use a trick called *artificial variables*. The idea is that, since we know how to improve a feasible solution, we will imbed our problem in a larger (artificial) problem in which it is easy to find a feasible solution; then we will improve this solution. The larger artificial problem includes some additional nonnegative variables called artificial variables. In a typical program we might have 57 structural variables to which we've already added 36 slack variables, and in this case a few more variables won't really hurt very much, provided that we get rid of them before we finish the problem. These artificial variables have several different interpretations, which, as we shall see, lead to minor modifications in the algorithm.

7.1.1 Minimization of Infeasibility

To motivate the discussion, and also to introduce a new class of problems which can be usefully solved by linear programming techniques, we return to the question of solution of simultaneous equations, which was begun in Chapter 1. In the absence of degeneracy, n equations in n unknowns always have a unique solution, but one or more of the variables may be negative in this solution. If we add nonnegativity constraints on all the variables, as we frequently must do in practical problems, then there may be no feasible solution to the set of equations, and in this case we may wish to know how close we can get to a solution. For example, if we can satisfy all the equations except one, and miss that one by only one unit, that's better than failing to satisfy one of the equations by several units or failing to satisfy several equations by one unit each. A simple example is graphed in Figure 7-1. The constraints (the equations to be solved) are

$$x_1 + x_2 = 1$$
$$x_1 + 2x_2 = 4$$

with the additional constraints

$$x_1 \geq 0 \qquad x_2 \geq 0$$

Without the nonnegativity constraints these equations have a unique solution: $x_1 = -2$, $x_2 = 3$. As the figure makes clear, there is no solution within the nonnegative region, but at least we're better off if we go to $x_1 = 0$ and $x_2 = 1$ (or 2, or somewhere in between), than if we go anywhere else on the diagram. As you can see, this problem is reminiscent of linear programming problems; the optimal solutions will occur at vertices, and if we set up some sort of a penalty function (or cost function) which measures the distances of our solution point

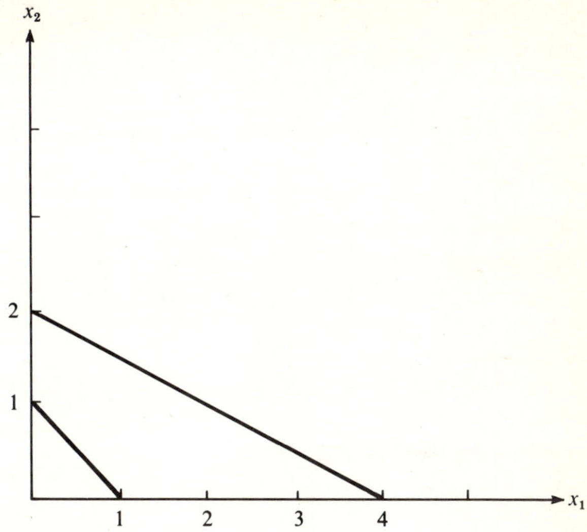

FIGURE 7-1
Simultaneous equations with no non-negative solution.

from the constraint lines, we begin to see how this might be modeled as a minimization of this linear cost function subject to appropriate linear constraints. More specifically, what we shall minimize is the total infeasibility of the problem, where total infeasibility is the sum of a set of component infeasibilities, and each component infeasibility measures the distance of our solution point from one of the constraint lines.

In order to do this, we add nonnegative artificial variables as follows:

$$\begin{aligned} x_1 + x_2 + a_1 \quad &= 1 \\ x_1 + 2x_2 + a_2 &= 4 \\ \text{Minimize:} \qquad a_1 + a_2 \quad &= f \end{aligned}$$

where the last line is called the *infeasibility form,* and f is total infeasibility. This is clearly a linear programming problem in feasible canonical form (at least the first two lines are in feasible canonical form), as we can see from calling a_1, x_3, calling a_2, x_4, and detaching the coefficients, in which case it becomes

$$\begin{array}{lccccc} & 1 & 1 & 1 & 0 & 1 \\ & 1 & 2 & 0 & 1 & 4 \\ \text{Minimize:} & 0 & 0 & 1 & 1 & f \end{array} \qquad (7\text{-}1)$$

If we let the nonbasic variables, x_1 and x_2, go to 0, we have the immediate solution $a_1 = x_3 = 1$, $a_2 = x_4 = 4$, $f = a_1 + a_2 = 5$. From Figure 7-1 you can see that this solution is at the origin, and clearly after we put the problem into the proper form, one iteration will take us up to the point $x_1 = 0$, $x_2 = 1$, where we have

$a_1 = 0$, $a_2 = 2$, $f = 2$, which minimizes the total infeasibility and is therefore an optimal solution.

Don't worry about the fact that $a_2 = 2$; this is not a direct measure of distance on the diagram, because it is scale-dependent. If we had written the second constraint equation in the form $\frac{1}{2}x_1 + x_2 = 2$ and had solved this problem instead, we would have ended up with everything the same except $a_2 = f = 1$.

Of course in this case there was no feasible solution in nonnegative variables, but suppose there had been. Then the minimum infeasibility of the problem would have been 0, and the linear programming method would have found it. That is, this is an alternative method for solving simultaneous equations, which is applicable in those cases where we are only interested in (or we happen to be sure that the solution will be in) nonnegative variables. It is not especially recommended as a substitute for the usual methods (for example, Gaussian elimination), but it does give us the key to how we will solve linear programming problems when we cannot easily obtain a feasible canonical form. We will add artificial variables until we do have a feasible canonical form; we will then minimize the infeasibility of the problem. When we have discovered this minimum infeasibility, either it will be positive or it will be 0 (it cannot be negative because it is the sum of nonnegative variables). If it is positive, then we know that the problem is infeasible, and we have obtained a solution which is as close as possible. If the minimum infeasibility turns out to be 0, we will know that the problem is feasible; but more important, we will have arrived at a solution in which each of the artificial variables is 0 (because they are all nonnegative and their sum is 0), and they may therefore be dropped. It follows that we have arrived at a feasible solution to our original problem, which we can improve if necessary to obtain an optimal feasible solution.

Because the C-vector of (7-1) is not in canonical form, we must, before proceeding any further, introduce subroutine D of our simplex algorithm, which tells us how to convert the C-vector into canonical form if the rest of the problem is in feasible canonical form. This simply involves generating 0s in the C-vector under each of the basic variables, and this is a very easy thing to do, because each basic variable has a coefficient of 1 in one row and 0s in all other rows. We shall demonstrate the procedure in equations (7-1). In these equations, the basic variables are x_3 and x_4, and we notice that c_3 is equal to 1 and c_4 is equal to 1. We start with c_3. To reduce this to 0 we simply have to subtract the first row from the cost row, which yields a new cost row:

$$-1 \quad -1 \quad 0 \quad 1 \qquad f-1$$

Now to change c_4 to 0 we simply subtract the second row from this new cost row, yielding

$$-2 \quad -3 \quad 0 \quad 0 \qquad f-5$$

If we append this new cost line to the A-matrix and B-vector of (7-1), we will have a problem ready for step 5 of the simplex algorithm:

$$\begin{array}{rrrrr} 1 & 1 & 1 & 0 & 1 \\ 1 & 2 & 0 & 1 & 4 \\ -2 & -3 & 0 & 0 & f-5 \end{array}$$

Clearly you will iterate with column 2 as the incoming variable; the optimal solution will be obtained in one iteration, namely,

$$x_1 = 0 \qquad x_2 = 1 \qquad a_1 = 0 \qquad a_2 = 2 \qquad f = 2$$

We now present this formally as

Subroutine D. To put the C-vector into canonical form if the A-matrix is in canonical form

Step D1 Locate a basic variable (call it column j) in which $c_j \neq 0$. Locate the row which has a one in column j. Multiply that row by c_j, subtract it from the cost row, and call this difference the cost row (discard the old cost row). Go to step 4 of the simplex algorithm (that is, repeat this step as often as necessary).

This may seem almost too simple, but when the C-vector is the initial infeasibility form it is even simpler than that, because each of the c_j's under the artificial variables has a coefficient of 1. Thus, you only have to add up all those rows in which artificial variables appear and subtract the sum from the infeasibility form; what remains is the cost row in canonical form (you discard the original infeasibility form).

7.1.2 The Two-Phase Method

In the kinds of linear programming problems to which we shall apply artificial variables, the first question to be answered is whether any feasible solution exists; if the answer to that is yes, we want to find some feasible solution—any basic feasible solution will do—in order to get started. In both cases, the original objective function is irrelevant; the question of feasibility depends only on the constraints. Therefore we discard (tentatively) the objective function, introduce artificial variables, and add a new objective function (the infeasibility form, f) which we proceed to minimize. When we have finished this first phase of the problem, the minimum infeasibility will be either positive (the problem is infeasible) or zero (we have arrived at a basic feasible solution to the original problem). In the latter case we put back the original objective function, and in the second phase of the problem we minimize that.

The algorithm for doing all this constitutes subroutine B of the simplex algorithm, which we are ready at long last to present. We will then be prepared (at least in theory) to handle any linear programming problem. We present the subroutine first and then give an example of it.

Subroutine B. *For putting a linear programming problem in standard form into canonical form (two-phase method)*

Step B1 Delete the cost function. Set $i = 0$. Go to step B2.

Step B2 Add one to i. If $i = m + 1$, go to step B5; if $i \leq m$, go to step B3.

Step B3 If the ith row contains a suitable pivot element (that is, if there is in the ith row a 1 in a column where all other coefficients are 0), go to step B2; if not go to step B4.

Step B4 Add an artificial variable whose coefficients are 1 in the ith row and the new cost row, and 0 in all other rows. Go to step B2.

Step B5 Fill out the new cost row with 0s under the A-matrix and the symbol f under the B-vector. Now minimize f by the simplex algorithm (starting with step 4, and hence subroutine D). If the minimum value of f is zero, go to step B7; if it is greater than zero, go to step B6.

Step B6 The problem is infeasible. Stop.

Step B7 Delete all artificial variables and the cost function. Add the original cost function. Go to step 4 (and hence subroutine D) of the simplex algorithm.

The idea of this algorithm is to create a canonical form by adding columns of 1s and 0s as necessary. In step B3 you are told not to add such a column if it is not necessary. The rows where it is not necessary are usually those in which the original constraint was of the less-than-or-equal-to type, and the slack variable was therefore introduced with a $+1$ coefficient and created the right kind of column. If a constraint had been of the equals or of the greater-than-or-equal-to variety, then there would have been no slack variable, or there would have been a slack variable with a coefficient of -1; in either case we will usually add an artificial variable.

At the end of this subroutine we go back to step 4 of the simplex algorithm. You may notice that we skipped step 3, and with it subroutine C. For the sake of completeness, we present it here:

Subroutine C. *To be used when the canonical form of a linear programming problem is infeasible (some $b_i < 0$)*

Step C1 If the canonical form is "optimal" and if you like the dual simplex algorithm, solve the problem by that method; otherwise, multiply each infeasible constraint equation (that is, each row with a negative b) by -1, and go to step 2 of the simplex algorithm.

In other words, you can always get a positive B-vector and then solve by artificial variables. But if the problem is amenable to solution by the dual simplex algorithm, that's usually a better way to do it.

Throwing out the cost function in step B1 is of special interest. We will, of course, put it back later, but at the moment our question is only whether or not the given problem is feasible. In answering this question, we are not interested in the nature of the original objective function; that is, whatever it was we would like to maximize or minimize is superflous. Thus the way this problem works is the following: Take out the cost function, create a canonical form by addition of artificial variables, set up a new objective function which is the total infeasibility of the problem, and minimize that infeasibility. The resultant minimum infeasibility will be either positive (proving the problem infeasible), or it will be 0. If it is 0, then you will immediately drop all artificial variables, throw out the cost function (which at this point will simply be a row of 0s, which is obviously of no interest anyhow), and put back the original cost function. It will now be necessary to apply subroutine D once more to put the original cost function into canonical form, and finally the original problem can be solved. You can thus see why this technique is called the *two-phase method*. In the first phase we work with the infeasibility form and minimize that; in the second phase we minimize the original objective function.

After we have inserted the artificial variables and put the infeasibility function into canonical form, we will have the same number of rows (m) as before and a few more columns. There will be m basic variables; each of the artificial variables will be basic, and the remaining basic variables (if any) will generally be those slack variables where we originally had a less-than-or-equal-to constraint. Now if the problem is feasible, then we must eventually drive each of the artificial variables to 0 before we complete Phase One; and since only one basic variable becomes nonbasic in each iteration, it will require at least as many iterations as there are artificial variables to complete Phase One. Since the simplex algorithm is not always perfectly efficient, it may require more than this number of iterations (that is, in some iterations one of the structural or slack variables may go from basic to nonbasic). However, there's one consolation: Once an artificial variable has become nonbasic, it will never again become basic. The reason for this is made obvious in the next section. Hence we have a shortcut, which can be introduced into the algorithm; namely, as soon as an artificial variable has become nonbasic, its column may be dropped, and no further computations need be made on it. What this means is that you never do any calculations in a column for an artificial variable: As long as it is basic, it's all 1s and 0s which you can copy; as soon as it becomes nonbasic, you can drop it. Thus, the artificial-variable method is much easier than it first seems.

It is possible for an artificial variable to become zero without becoming nonbasic. Obviously, this occurs only when the problem is degenerate. If this happens, it may be worth your while to find out why, because it probably means that you wrote some redundant constraints. You cannot actually drop the artificial variable until it becomes nonbasic.

We now give an example of the application of this algorithm, but since such problems tend to become rather large and tedious, we do not present it in full detail. Assume that the problem is

$$\begin{aligned} x_1 + 2x_2 &\leq 12 \\ 2x_1 + x_2 &\leq 10 \\ x_1 + 3x_2 &\geq 4 \\ 3x_1 + x_2 &\geq 5 \end{aligned} \tag{7-2}$$

$$\text{Minimize:} \qquad x_1 + x_2 = z$$

Now if you're clever you may notice that, since this is a minimization problem, the first two constraints can't possibly have anything to do with the optimum solution; you could throw these out, solve the remaining problem quickly by the dual simplex method, and then (just to make sure you were right) check to be sure that the final solution did not violate the first two constraints. However, real problems may be much too complex for such intuitive analysis, and so we proceed with a full formal solution of the problem.

Adding slack variables to put the problem into standard form, and detaching coefficients,

$$\begin{array}{rrrrrrr} 1 & 2 & 1 & 0 & 0 & 0 & 12 \\ 2 & 1 & 0 & 1 & 0 & 0 & 10 \\ 1 & 3 & 0 & 0 & -1 & 0 & 4 \\ 3 & 1 & 0 & 0 & 0 & -1 & 5 \\ 1 & 1 & 0 & 0 & 0 & 0 & z \end{array} \tag{7-3}$$

Now we go to Phase One of the two-phase procedure; in accordance with subroutine B we delete the old cost function, insert artificial variables with 1s in rows three and four, and add a new cost function, as follows:

$$\begin{array}{lrrrrrrrrr} x_3 & 1 & 2 & 1 & 0 & 0 & 0 & 0 & 0 & 12 \\ x_4 & 2 & 1 & 0 & 1 & 0 & 0 & 0 & 0 & 10 \\ x_7 & 1 & 3 & 0 & 0 & -1 & 0 & 1 & 0 & 4 \\ x_8 & 3 & 1 & 0 & 0 & 0 & -1 & 0 & 1 & 5 \\ & 0 & 0 & 0 & 0 & 0 & 0 & 1 & 1 & f \end{array} \tag{7-4}$$

Subtracting rows three and four from the cost function in order to put it in canonical form, we arrive at

$$\begin{array}{lrrrrrrrrr} x_3 & 1 & 2 & 1 & 0 & 0 & 0 & 0 & 0 & 12 \\ x_4 & 2 & 1 & 0 & 1 & 0 & 0 & 0 & 0 & 10 \\ x_7 & 1 & 3 & 0 & 0 & -1 & 0 & 1 & 0 & 4 \\ x_8 & 3 & 1 & 0 & 0 & 0 & -1 & 0 & 1 & 5 \\ & -4 & -4 & 0 & 0 & 1 & 1 & 0 & 0 & f-9 \end{array}$$

Now because there are two artificial variables, it's clear that it's going to take us at least two iterations to drive them both to 0. It turns out that in this simple problem it takes only two iterations, at the end of which we will have reached

x_2	0	1	0	0	$-\frac{3}{8}$	$\frac{1}{8}$	$\frac{3}{8}$	$-\frac{1}{8}$	$\frac{7}{8}$
x_3	0	0	1	0	$\frac{5}{8}$	$\frac{1}{8}$	$-\frac{3}{8}$	$-\frac{1}{8}$	$\frac{71}{8}$
x_4	0	0	0	1	$\frac{1}{8}$	$\frac{5}{8}$	$-\frac{1}{8}$	$-\frac{5}{8}$	$\frac{51}{8}$
x_1	1	0	0	0	$\frac{1}{8}$	$-\frac{3}{8}$	$-\frac{1}{8}$	$\frac{3}{8}$	$\frac{11}{8}$
	0	0	0	0	0	0	1	1	$f-0$

Actually you would not have to do quite so much work, because in the first iteration, when x_8 went to 0, you would have dropped it and not bothered to calculate its column, and similarly with x_7 in the second iteration. However we thought you might be interested to see what these columns look like. The final row of the tableau simply says that $a_1 + a_2 = f$, which tells us nothing we didn't know already, and is therefore no longer useful to us. Dropping this row, dropping the columns corresponding to the two artificial variables, and putting back the original cost function, we arrive at

x_2	0	1	0	0	$-\frac{3}{8}$	$\frac{1}{8}$	$\frac{7}{8}$
x_3	0	0	1	0	$\frac{5}{8}$	$\frac{1}{8}$	$\frac{71}{8}$
x_4	0	0	0	1	$\frac{1}{8}$	$\frac{5}{8}$	$\frac{51}{8}$
x_1	1	0	0	0	$\frac{1}{8}$	$-\frac{3}{8}$	$\frac{11}{8}$
	1	1	0	0	0	0	z

Now we need one more application of subroutine D in order to convert this cost function into canonical form. We get it by subtracting the first and fourth rows from the cost row, yielding

x_2	0	1	0	0	$-\frac{3}{8}$	$\frac{1}{8}$	$\frac{7}{8}$
x_3	0	0	1	0	$\frac{5}{8}$	$\frac{1}{8}$	$\frac{71}{8}$
x_4	0	0	0	1	$\frac{1}{8}$	$\frac{5}{8}$	$\frac{51}{8}$
x_1	1	0	0	0	$\frac{1}{8}$	$-\frac{5}{8}$	$\frac{11}{8}$
	0	0	0	0	$\frac{1}{4}$	$\frac{1}{4}$	$z-\frac{9}{4}$

To our surprise and pleasure, the problem is already in optimal form, and we do not have to go through Phase Two of the optimization procedure. This will not generally happen in large problems, but it will frequently happen that the canonical form we arrive at at the end of Phase One represents a pretty good solution and requires only a few more iterations to complete. You may note in this solution that, as we predicted, the first two slack variables are positive; that is, the first two constraints are inactive.

7.1.3 The Big-M Method

We now present an alternative method of solving with artificial variables. Let's go back to the problem (7-2) and assume that it's a diet problem, where x_1 is pounds of meat and x_2 is bushels of potatoes. The first two constraints refer to two poisons in the foodstuffs, say, arsenic and mercury; the last two constraints refer to dietary constituents, say, protein and vitamins. As before, we go to the standard form (7-3), where x_3 and x_4 are the amounts by which we are within our permissible limits of poisons, and x_5 and x_6 are the extra amounts of protein and vitamins which we get, over and above our requirements. Now what we need to solve this problem is some sort of a first feasible solution, and we shall obtain this by introducing two imaginary foodstuffs. The first, of which we shall purchase x_7 pounds, supplies precisely one unit of protein, none of anything else, and costs 1 million dollars for each pound. Similarly, the second provides one unit of vitamins for each pound of x_8 and also costs 1 million dollars a pound. Adding these to the problem (7-3) without deleting the cost function, we get

$$\begin{array}{ccccccccccc} x_3 & 1 & 2 & 1 & 0 & 0 & 0 & 0 & 0 & & 12 \\ x_4 & 2 & 1 & 0 & 1 & 0 & 0 & 0 & 0 & & 10 \\ x_7 & 1 & 3 & 0 & 0 & -1 & 0 & 1 & 0 & & 4 \\ x_8 & 3 & 1 & 0 & 0 & 0 & -1 & 0 & 1 & & 5 \\ & 1 & 1 & 0 & 0 & 0 & 0 & M & M & & z \end{array} \qquad (7\text{-}5)$$

The first four rows of (7-5) are identical to the first four rows of (7-4), but the cost function is different. The cost function of (7-5) is the sum of the cost function (7-3), plus a modified cost function of (7-4), where we have charged M for each pound of x_7 or x_8 instead of 1. We have tentatively called M 1 million dollars, which is reasonable if all the other costs are around \$1 or \$2 each; but if we were doing something for the federal government, you might have to make these imaginary materials cost much more than 1 million dollars each. We symbolize this by M, where M simply represents a very large number, just as earlier we symbolized a very small number by ϵ. If you would like to, you may think of $\epsilon = 1/M$.

Now the problem of (7-5) can be converted into canonical form by subroutine D, and we can proceed by the conventional simplex method to solve the problem. It turns out that we will do exactly the same things that we did before, but we will not need two phases. Before, we made changes in the A-matrix and B-vector during Phase One, and then caught up on all the corresponding changes in the C-vector before starting Phase Two. Here, we do both at once. The result is a little messy. For example, applying subroutine D to (7-5) yields, for the first coefficient in the C-vector, $c_1 = -4M + 1$. Eventually you will have to multiply a whole row by $-4M + 1$. Thus, this alternative form of the algorithm has fewer steps, but each step has more work. The formal description of this algorithm is left as an exercise (Problem 7.1-1).

As before, each artificial variable should be dropped as soon as it becomes nonbasic; eventually all artificial variables will be dropped and all M's will disappear out of the cost function (if the original problem is feasible). In all other respects the algorithm is identical; it does give an interesting new interpretation as to what the artificial variables mean. Each artificial variable is now an imaginary decision variable which satisfies precisely one constraint without affecting any others and which is terribly expensive. Since the simplex algorithm always finds the minimum-cost solution, it will always find a solution which does not include any of these terribly expensive artificial variables if such a solution exists. If no feasible solution is possible without using some of the artificial variables, then the original problem was, of course, infeasible. You can also see why a nonbasic artificial variable never comes back in (and therefore why you can drop it). If we have a feasible solution not including some imaginary (and expensive) material, we will obviously not want to use such material. And you remember that the simplex method always goes from one feasible solution to another.

In summary, artificial variables constitute a (legal) method of cheating: If you don't have enough columns of ones and zeros to make a canonical form, just add such columns until you do. Then solve, ensuring somehow that these cheaty columns aren't around at the end.

Problems

7.1-1 Write subroutine B for the big-*M* method.

7.1-2 Do Problem 5-6 by the two-phase method (that is, use the simplex method with subroutine B, rather than the dual simplex). This is tedious. Phase One will take three iterations, and Phase Two one more. Your worst fractions will be twelfths; but all fractions will eventually disappear, and your final tableau will (of course) be the same as before. We hope at least to convince you that the dual simplex algorithm is pretty neat!

7.1-3 Do Problem 5-7 by the big-*M* method.

7.1-4 Add to Problem 4-5 the constraint $x_1 + x_2 + x_3 \geq 6$.

(*a*) Use the two-phase method to prove that this problem is insoluble. *Hint:* In the first iteration, you will have a tie for most negative c_j. It is suggested that you not use the left-hand tie-breaking rule in this case.

(*b*) Show that you just wasted a lot of time, because it is obvious by inspection that this expanded problem has no feasible solution.

7.1-5 Minimize: $2x_1 + 5x_2 + 7x_3$

Subject to:

$$\begin{aligned} 3x_1 + 2x_2 + x_3 &\leq 100 \\ x_1 + 4x_2 + 2x_3 &\leq 100 \\ x_1 + x_2 + 3x_3 &\geq 100 \\ x_1 \geq 0 \qquad x_2 \geq 0 \qquad & x_3 \geq 0 \end{aligned}$$

Extra Drill Problems

You can solve any problem in Chapter 5 by the two-phase or big-*M* method.

7.2 SENSITIVITY ANALYSIS AGAIN

In Section 6.8 we discussed sensitivity analysis and postoptimality analysis for the transportation problem. We now study these questions as they apply to the general linear programming problem, but there are no new principles. We already know everything we have to know to do the postoptimality analysis, and it's merely a question of giving some examples of how it works.

The things that may change are: addition of a new variable or deletion of an old one, addition of a new constraint or deletion of an old one, change of a b (requirement), change of a c (cost), and change of an a (activity coefficient). If we change a parameter (that is, if we change an a, b, or c), we can ask several kinds of questions, some of which may be easy and others hard. One question is: If we make a small change in the value of a particular parameter, what happens? If the answer to that question is "nothing" (or "not very much," because the same set of basic variables constitute the optimal feasible solution), then the second question might be: Over what range of change does the present solution remain optimal? Or if the basic variables do change, the new question might be: Which variable goes in and which goes out? And in either case, we might want to know the new values of the variables and objective function. Finally there will be two different kinds of calculations, again one of which may be easy and the other hard, depending upon whether the particular a or c is in a column which is basic or nonbasic and whether the particular a or b is in a row corresponding to an active or inactive constraint.

Do you recognize that the two statements in the last sentence are equivalent? Consider a particular row of the (primal) problem in which the constraint is inactive; that is, the slack variable for this row comes out positive in the final solution, and a small change of the corresponding b will not affect this solution. As we explained in great detail in Section 5.4, the shadow price of this constraint is 0, which is to say that in the dual problem the corresponding dual variable is 0. Thus this same b, which is considered to be a requirement in the primal problem, may be considered to be a cost in the dual problem; if the corresponding variable of that dual problem is nonbasic, then it is not used, and a small change in its cost will not affect how much of it we use.

This gives us a hint as to some of the tricks that can be used in postoptimality analysis. Remember that when we have finished solving the problem, the solution is displayed for both the primal and dual problems. To refresh your memory, the solution to the basic variables of the primal problem is displayed in the B-vector, and for the nonbasic variables the solution is implicitly zero. The optimal values of the dual structural variables are displayed under the slack variables in the final tableau, and the optimal values of the dual slack variables are displayed under the structural variables in the final tableau. The optimal z-value is displayed in the lower right-hand corner, and the optimal w-value is equal to it. It follows that we are free to answer any question by looking at either the dual or the primal problem, whichever is more convenient.

For example, suppose we have a solution and somebody comes up with a new variable that we hadn't thought about before (a new product that might be made or a new foodstuff that might go into our diet). We are given the activity coefficients and the cost (profit) for this new variable and asked whether or not it should come into the optimal solution. The easiest way to answer this question is to look at the dual problem; this column becomes a row of the dual problem, which is therefore a constraint, and it is very easy to determine whether the optimal values of the dual problem satisfy this constraint. If they do, then we've answered the question: Don't change anything; if they don't, then we have to iterate again. This is tedious, since you must insert the new column in the first tableau and then operate on it through each successive tableau before you can perform this new iteration. Fortunately, there is a short cut for doing all this, which is described in Section 7.3.

7.2.1 Change of b (Requirement)

We repeat here the example problem which we worked on in Sections 4.6 and 5.4:

Maximize: $2x_1 + 3x_2 = z$

Subject to:

$$\begin{aligned} 2x_1 + x_2 &\le 5 \\ x_1 + x_2 &\le 3 \\ x_1 + 2x_2 &\le 5 \\ x_1 \ge 0 \quad x_2 &\ge 0 \end{aligned} \tag{4-6}$$

The final tableau is

	x_1	x_2	s_1	s_2	s_3	b
s_1	0	0	1	−3	1	1
x_1	1	0	0	2	−1	1
x_2	0	1	0	−1	1	2
	0	0	0	1	1	$-z+8$

(5-6)

Look first at b_1—that is, the number 5 on the right-hand side of the first row of (4-6); what happens if we increase it? Obviously nothing. From the final tableau it's clear that we have one unit of this resource left over, and so additional units of this resource cannot possibly affect the situation. What happens if b_1 decreases? Obviously, since s_1 is 1 in the final tableau, nothing happens until b_1 is reduced to 4. At this point we have a degeneracy; and if b_1 becomes less than 4 we have a different optimal feasible solution. To find out exactly what this solution is requires a new iteration (dual simplex).

Now let's look at b_2, which was originally 3. Since $s_2 = 0$ in the final solution, any change in this number will affect the final solution. The change in cost is given by the opportunity cost of s_2, namely, 1. That is, if we increase b_2 from 3 to 3.01, we will increase the profit from 8 to 8.01, and vice versa. This ratio of change in profit to change in resource remains the same only as long as the present basic solution remains optimal, that is, only as long as x_1, x_2, and s_1 remain the basic variables. Clearly if b_2 became extremely large, say, 100, we would not be able to use up all this resource, and s_2 would become basic. Clearly, if b_2 became extremely small, say, $\frac{1}{100}$, we would not be able to use up all the third resource, and s_3 would become basic.

Therefore, the next question for us to investigate is: Over what range of values for b_2 does the present solution remain optimal?

For this purpose we will add the very small number ε (epsilon) to b_2, but this time we will let ε be either positive or negative, and we will ask the quantitative question: How small is "very small" for this particular problem? A remarkable property of the simplex algorithm enables us to answer this question rather easily. Suppose you write the standard form corresponding to (4-6), changing b_2 from 3 to $3 + \varepsilon$ (please do this on a piece of scratch paper). Now note that the coefficient of s_2 is exactly the same as the coefficient of epsilon, namely, 1. Now perform the first iteration on this tableau; notice that whatever you do to s_2, you must do exactly the same thing to ε. This continues to be true at each succeeding iteration, and it follows that in the final tableau the coefficients of epsilon must be the same as the coefficients of s_2 throughout the tableau. It follows that we can rewrite by inspection the final tableau, since the coefficients of ε must be the same as the coefficients of s_2. That is, the new final tableau becomes

	x_1	x_2	s_1	s_2	s_3	b	
s_1	0	0	1	-3	1	$1 - 3\varepsilon$	
x_1	1	0	0	2	-1	$1 + 2\varepsilon$	(7-6)
x_2	0	1	0	-1	1	$2 - \varepsilon$	
	0	0	0	1	1	$-z + 8 + \varepsilon$	

Note first that $z = 8 + \varepsilon$, which is a quantitative restatement of our assertion about opportunity costs. Next note that for this solution to be feasible each of the numbers on the right-hand side of (7-6) must be nonnegative. Thus, the first row tells us that $1 - 3\varepsilon \geq 0$, which implies $1 \geq 3\varepsilon$, or equivalently, $\varepsilon \leq \frac{1}{3}$. The second row tells us that $1 + 2\varepsilon \geq 0$, or $\varepsilon \geq -\frac{1}{2}$. And the third row tells us that $2 - \varepsilon \geq 0$, or $\varepsilon \leq 2$. In order to satisfy all three of these inequalities, it follows that we must have $-\frac{1}{2} \leq \varepsilon \leq \frac{1}{3}$ and finally, therefore, that the present set of basic variables remains the optimal set for $2\frac{1}{2} \leq b_2 \leq 3\frac{1}{3}$. Usually this is a sufficient answer to the question. However, it might turn out that b_2 was equal to 2, and we then had to find the new optimal feasible solution. But $b_2 = 2$ implies $\varepsilon = -1$, which we may substitute into (7-6). This final tableau is now in an "optimal" canonical form to which we can apply the dual simplex algorithm and quickly find the new solution.

7.2.2 Change of c (Cost)

If one of the costs in the original problem changes, you may (if you wish) attack the dual problem exactly as above. However, the theory of duality also tells us that we may perform an exactly analogous function working directly with the primal tableau. Obviously if the c corresponds to a nonbasic variable, then changing the cost in one direction can never affect the problem; changing the cost in the other direction will have an effect only if this change is sufficiently great to drive the corresponding c in the final tableau to 0 or beyond. Again notice that if we change the cost in the original standard form by epsilon, the cost in the final tableau will change by exactly epsilon (remember that if it is a maximization problem, adding ε to the cost means adding $-\varepsilon$ to the standard form).

The more difficult (and more interesting) situation is where the variable is basic. Suppose you change c_1 in (4-6) from 2 to $2 + \varepsilon$; then c_1 in (5-6) will change from 0 to $-\varepsilon$. You must now apply subroutine D to get the 0 back there. (Or you can iterate once more, which turns out to be exactly equivalent.) When you do, you will have $c_4 = 1 + 2\varepsilon$ and $c_5 = 1 - \varepsilon$; both of these numbers must be positive if the solution is to remain optimal, and so it follows that we must have $-\frac{1}{2} < \varepsilon < 1$, which says that the present basic solution remains the unique optimal basic solution for $1\frac{1}{2} < c_1 < 3$.

In summary, any change in a b_i or a c_j may result in a problem which can be solved by inspection or with very little work. At worst, it will result in an almost-solved problem to which the simplex or dual simplex algorithm may be applied, as appropriate. Note that changing c_j can affect the optimality, but not the feasibility, of a particular optimal feasible basic solution; changing a b_i can affect both.

7.2.3 Change of a (Activity Coefficient)

First let us point out that every a_{ij} in every tableau, and in particular in the final tableau, is meaningful. Clearly only the a_{ij}'s for the nonbasic variables are interesting (because all the others are 1s and 0s). Consider then, for example, in (5-6) the coefficient a_{24}; that is, the coefficient of s_2 in the second row, which is equal to 2. It tells you exactly the tradeoff between the variable of its column (in this case s_2) and the variable of its row (in this case x_1). That is, it tells you that if you wish to bring in one unit of s_2, you have to give up two units of x_1. Examining the rest of the s_2-column, we find that if we want to bring in one unit of s_2 we have to give up -3 units of s_1 (that is, we will actually get three extra units of s_1), and we have to give up -1 unit of x_2 (that is, we will get an extra unit of x_2). The final row tells us that if we do bring in an extra unit of s_2, it's going to cost us \$1. All the above statements, as you know, do not necessarily apply to a whole unit; they are valid only over that region in which the present basic solution remains optimal—that is, until we have moved to the next adjacent vertex and there is a change in the set of basic variables. Thus, this use of the a_{ij}'s makes the complete sensitivity analysis very simple as long as the same set of variables remains basic.

Note how easy this makes certain kinds of postoptimality analysis even in very complicated problems. Take the final tableau presented in Problem 4-6 (page 140) for example. We are not now producing any x_4. Suppose it is decided that we should produce some (either because c_4 changed, or for reasons outside the model). Obviously, for each unit of x_4, it will cost us two units of money. What else happens? We give up two units of x_2 and get one extra unit of x_3 and x_5. All other variables remain zero. Obviously since $x_2 = 1$ in the present solution, you can only bring in one-half unit of x_4 before this argument becomes invalid.

A little thought should convince you that if a variable is nonbasic in the final solution, then a change in one of the original a_{ij}'s for that j can affect only the optimality (not the feasibility) of that solution, and even that only if the a is changed in a particular direction and by at least a certain amount. To see this, we need a problem at least as complicated as (5-1), page 151, whose final tableau is presented in (5-4). Note first that it does not matter by which algorithm this final tableau was achieved. Note next that y_1 is nonbasic in the final solution. Now if $a_{11} < 2$ (that is, if each unit of meat supplied less than two units of protein) or if $a_{21} < 1$, there would be even less reason to use y_1. On the other hand, if $a_{21} = 2.5$, then a feasible solution would be $y_1 = 0.4$, $y_2 = 2$, $y_3 = 0$, $w = 8$, which is alternate optimal, and for $a_{21} > 2.5$ we could find a cheaper feasible solution ($y_1 < 0.4$, $y_2 = 2$, $y_3 = 0$, $w < 8$). Similarly, if $a_{21} = 1$ and a_{11} were increased to 3, we would have an alternate-optimal basic feasible solution: $y_1 = \frac{1}{5}$, $y_2 = 0$, $y_3 = \frac{7}{5}$; and for $a_1 > 3$ we could find better solutions.

If we change an a_{ij} for a variable which is basic in the final tableau, things become even messier, because now either feasibility or optimality may be affected. But even in the simpler (nonbasic) case, while we have presented a solution, we have not given you a general method of finding a solution, because it is just too complicated to be worthwhile unless one uses matrix algebra. The necessary algebra is presented in the following section. Whether or not you study that section, we hope you have realized by now that the reason for introducing fancy mathematics is not to make simple things complicated but to make complicated things simple. Our matrix notation and the other mathematical tricks which we have presented, supplemented where necessary by the rules presented in Section 7.3 (which are derived from matrix algebra), make it possible for us not only to solve very complicated problems but also to obtain considerable insights into those problems (as by postoptimality analysis) with very little work.

Problems

7.2-1 In Problem 5-3 we displayed the final tableau for the problem of Section 4.11.1. The present solution remains optimal over what range of prices for

(*a*) Pure-beef frankfurters?

(*b*) Hot dogs?

and over what range of availabilities for

(*c*) Beef trimmings?

(*d*) By-products?

(*e*) The president decides to make 1000 pounds of hot dogs anyhow to maintain the company image. Now,

(i) How much money do you make?

(ii) How much of each product do you make?

(iii) How much of each constituent is left over?

Hint: In answering (ii) and (iii) use the a_{ij}'s.

7.2-2 In Problem 4-6,

(*a*) What would happen if c_4 were increased from 3 to 4? To 5? To 6?

(*b*) What would happen if c_3 were changed?

(*c*) The b_1 given in the original problem was 7. Suppose it turned out to be only 6 after all. Would this affect the solution? If so, what would the new solution be?

(*d*) Suppose b_1 were reduced from 7 to 5; why is this situation different from that of part (c)?

(*e*) What would the new solution be?

In each part, only one parameter is changed; thus, $c_4 = 3$ for parts (*b*) through (*e*).

7.2-3 (*a*) Solve the problem of Problem 5-2; that is,

$$\begin{array}{ll} \text{Maximize:} & 4x_1 + 3x_2 - x_3 \\ \text{Subject to:} & x_1 + 3x_2 - x_3 \geq 12 \\ & 2x_1 - x_2 + x_3 \leq 10 \\ & x_1 + x_2 = x_3 + 5 \\ & x_1 \geq 0 \quad x_2 \geq 0 \quad x_3 \geq 0 \end{array}$$

Something peculiar will happen. When it does, explain it.

(*b*) Solve the dual of this problem, which is your answer to Problem 5-2*a*.

Extra Drill Problems

Choose any problem in Chapter 4. Choose a b or c and do the sensitivity analysis. Then change this b or c and re-solve the problem (or have a computer do it for you) to see if the sensitivity analysis was correct.

7.3 THE REVISED SIMPLEX METHOD AND DECOMPOSITION

Most of the work in carrying out simplex iterations consists of computing the a_{ij}'s (activity coefficients) in accordance with step E6 of subroutine E. We have pointed out that sometimes some of this work can be skipped; in particular, it is never necessary to compute any a_{ij}'s for artificial variables, and in connection with Problem 5-8 we pointed out that you don't really need to compute any a_{ij}'s in the last tableau once you know that it is the last tableau (unless you want to use them in postoptimality analysis). All we're interested in in the last tableau is the b's and c's. We need the a_{ij}'s only if we have to iterate again; but now note that in each iteration, the only column of a_{ij}'s in which we are really interested is the column for the incoming variable. All the other a_{ij}'s in any given tableau are used only for the purpose of computing the a_{ij}'s in the next tableau, and the only reason we compute those is in case they might turn out to be the incoming variable, in which case we'd need them to determine which row to circle.

Now suppose that we had a product-mix-type problem which (to take an extreme example) had two rows and 50 columns. We might anticipate that two or three iterations would be required to reach the optimal feasible solution, and thus there are only two or three columns of a_{ij}'s that ever have to be computed. Nonetheless, the simplex algorithm (as we have learned it thus far) requires us to compute 50 columns of a_{ij}'s for each iteration. The revised simplex method is a way of avoiding this effort. Specifically, there is a method which allows us to do three things at each iteration: (1) compute each of the b_i, which constitute the present values of the basic variables; (2) compute each of the c_j, so that we may determine which is the incoming variable; (3) compute each of the a_{ij} in the one interesting column, namely, the column for the incoming variable. It turns out that we can do these things without computing any of the activity coefficients for any of the structural variables except the incoming variable; it is required, however, that we compute all the activity coefficients for all the slack variables.

Why we do what we do in the revised simplex method is rather easy to understand if you are familiar with matrix algebra and rather hard to understand if you are not. Since we assume that you are not, we're not going to try to explain it at all, but simply present the algorithm in "cookbook" fashion. This algorithm differs essentially only in step E6. We assume that you have in the old tableau all the b_i, all the c_j, and the a_{ij} for the slack variables and for the incoming variable; we will show how to compute these for the new tableau. For our notation we will assume that there were initially n structural variables and m constraints (m slack variables) so that the slack variables are $x_{n+1}, \ldots, x_{n+m}$. We also differentiate between a_{ij}, the coefficients in the "old" tableau, and a_{ij}^0, the coefficients of the initial tableau (where the slack variables were the basic variables), and correspondingly for c_j and c_j^0; that is, the superscript zero identifies elements from the very first canonical form. Proceed as follows: Compute the last $m + 1$ columns exactly as before by Gaussian elimination; that is, compute the m rows of a_{ij} and b_i and the bottom row of c_j and z (for the slack variables and b column) by step E6. Now for each remaining nonbasic variable, compute c_j as follows: Multiply c_{n+1} by a_{1j}^0, where c_{n+1} is the cost coefficient of the first slack variable in the old tableau, and a_{1j}^0 is the activity coefficient of the jth variable in the first row of the initial tableau; similarly, multiply c_{n+2} by a_{2j}^0; and so on for m such products. Add up these m products, and subtract the sum from c_j^0 (the c_j for the jth variable in the first tableau). This difference yields the c_j in the new tableau. The formula is

$$c_j = c_j^0 - (c_{n+1}a_{1j}^0 + c_{n+2}a_{2j}^0 + \cdots + c_{n+m}a_{mj}^0) = c_j^0 - \sum_{k=1}^{m} c_{n+k}a_{kj}^0 \qquad (7\text{-}7)$$

The quantity in parenthesis is often called z_j. Many books restrict the symbol c_j to what we have here called c_j^0; what we have here called c_j is called by such books $c_j - z_j$. Note that our awkward use of c_j and c_{ij} in Sections 6.5 and 6.7 could have been thus avoided; c_j would become $c_{ij} - z_{ij}$.

Now you will return to step 5; if there are negative c_j's, you will pick the most negative, and it is then necessary to determine the activity coefficients for this column, which we shall call column j. Proceed as follows: To compute the activity coefficient in the first row, multiply each of the activity coefficients in the jth column of the first tableau by the corresponding coefficients of the slack variables in the first row of the new tableau, and add the products. Specifically,

$$a_{1j} = a_{1j}^0 \times a_{1,n+1} + a_{2j}^0 \times a_{1,n+2} + \cdots + a_{mj}^0 \times a_{1,n+m}$$

In general, for the ith row,

$$a_{ij} = a_{1j}^0 \times a_{i,n+1} + a_{2j}^0 \times a_{i,n+2} + \cdots + a_{mj}^0 \times a_{i,n+m} = \sum_k a_{kj}^0 a_{i,n+k} \qquad (7\text{-}8)$$

Now go back to step E1.

The set of activity coefficients of the slack variables in any tableau constitutes an $m \times m$ matrix which is called the "inverse of the basis matrix"; so in case you see this phrase on a computer printout that is using the revised simplex method, you'll know what it means.

As we said, it's more trouble than it's worth to try to explain this. You may take our word for it; or, if you prefer, go back to any one of the simplex tableaux in this book and try it. For example you may wish to derive the circled column of (4-10) on page 123 from the four right-hand columns of (4-10) and the second column of (4-9). This is pretty trivial, because the inverse of the basis matrix is here mostly 1s and 0s. If you want to try one that's less trivial, then you might try deriving the third column of the tableau of Problem 5-3. You are asked to do this in Problem 7.3-1.

In case you do happen to know something about matrix algebra and the theory of linear vector spaces, we can make a little bit of sense out of the above. You remember that one way to invert a square matrix is to place an identity matrix on its right, and then perform Gaussian iterations to develop the identity matrix on the left, in which case the matrix on the right is now the inverse of the original matrix. It turns out that this is what we have been doing all along in the simplex. We started with the identity matrix on the right in the form of a set of slack variables or other canonical form. What is left in these last m columns at any time is precisely the inverse of the matrix of whatever elements were in those m columns of the original matrix which now correspond to the m basic variables. These m variables are called the *basis*, and what we have under the slack variables in any tableau is precisely the inverse of that basis matrix. It is called the basis because, in terms of linear vector spaces, that is precisely what it is: The m basic variables constitute linearly independent vectors in an m-dimensional vector space, and using these vectors as a basis, each of the remaining variables has been expressed as a linear combination of the basis vectors. Furthermore, the a_{ij}'s in the tableau are simply the appropriate coefficients for such linear combination. For this reason, the activity coefficients of the jth nonbasic vector can be appropriately obtained by multiplying the vector from the original tableau by the inverse of the basis matrix, which is precisely what we have done above.

We have indicated that the revised simplex method saves time if there are many more columns (structural variables) than rows (constraints) in the conventional form of the problem. It saves even more time if there are lots of zeros in the original matrix, since then many of the multiplications indicated in (7-7) and (7-8) do not need to be performed. Such *sparse matrices* (matrices with lots of zeros) are common in real problems because particular constraints may involve only a few variables; that is, not every variable competes for all resources. Equations (3-4) or (3-5), page 85, provide an example of a sparse matrix.

Suppose we had a real problem with 300 structural variables and 100 constraints (much larger problems are routinely solved in seconds or minutes on modern computers). Suppose further that the 37th constraint involved only x_1 through x_{20}, and the 38th only x_{21} through x_{40}. Then there would be 280 zeros in each of these rows (not counting all the zeros under the slack variables) in the first tableau. Now suppose that the 37th row became the pivot in the first iteration. Then the 38th row in the second tableau would probably have nonzero numbers in columns 1 through 40, and only 260 zeros. As you can see, the sparseness tends to decrease as we go through simplex iterations; but the revised simplex continues to work with the sparse first tableau.

Some interesting shortcuts are available with sparse matrices. For example, if there is a zero anywhere in the circled column, the entire row in which it lies may be copied without change into the next tableau (why?); and for every zero in the circled row (there will always be some), the entire column in which it lies may be copied into the next tableau, provided that you are not rearranging rows (why?). Whether it is worthwhile to modify the algorithm to take advantage of such shortcuts is a matter of choice.

If the matrix is sparse, it may be possible to *decompose* the big problem into several easier small problems. In particular, if a group of constraints applies to only a few of the variables, and these variables are absent from most of the remaining constraints, then this group of constraints can be deleted to form an easy subproblem. Eventually the solutions to the subproblems have to be put back together to get the solution to the original problem. This *decomposition algorithm* is too complicated for us, but it may be found in more advanced books on linear programming.

Problems

7.3-1 Use the revised simplex method to compute the x_3-column of Problem 5-3. You are to use the last four columns of this problem together with the third column of the original problem on page 134. Note that the first four rows of this problem (on page 134) are in canonical form but that the cost row is not; you must use the canonical cost form which is developed immediately below the problem statement.

7.3-2 Complete the formulation (Section 7.2.3) of how to handle changes in a_{ij}.
(*a*) What happens if a_{11} in equation (5-1) is changed from 2 to 2.5? To 3.5?
(*b*) What happens if a_{12} in equation (5-1) is changed from 1 to 0.5? To 2?

7.3-3. Solve Problem 4-9 by the revised simplex method.

7.4 SOME EXTENSIONS OF THE ASSIGNMENT AND TRANSPORTATION PROBLEMS

We have already looked at several modifications of the assignment and transportation problems, which require modifications of our algorithms or the development of brand-new algorithms. The assignation problem of Section 2.8 turned out to be just an assignment problem, but the three-dimensional assignment problem of Problem 6-15 turned out to be very different. In Problem 6-6 we saw a modification of the transportation problem whose effects we will study further in Section7.6, and in Problem 6-14 we saw another modification which we will further discuss in Problem 7.5-2; neither of these can be solved by the methods covered so far in this book.

As we said in Section 2.6, it's hard to tell when looking at a modification of a problem whether it makes it a little bit harder or much harder. In Section 7.4.1 we shall examine a modification of the assignment problem which makes it only a little bit harder; and in Sections 7.4.2 and 7.5 we shall examine modifications which make it enormously harder. In Sections 7.4.3 to 7.4.6 we shall examine four modifications of the transportation problem in which, again, some are hard and some are easy. In Section 7.4.7 we shall outline briefly a theoretical framework for handling many such modifications.

7.4.1 The Bottleneck Assignment Problem

We are to assign n workers to n jobs on a production line. Here the "costs" are the time required for the ith worker to do the jth job, and we wish to minimize time so that the production rate can be as high as possible. However, the line can move no faster than its slowest element, and so what we wish to minimize is not the sum of the times but the "bottleneck" time; that is, the time of the slowest worker on the line. It really doesn't matter whether the first job is done in 2 seconds or in 17 seconds; if the slowest job on the line takes 25 seconds, the whole line can move no faster than that.

Thus, for the formal statement of the problem, the constraints are exactly as in equations (2-1), but the objective function is different; it becomes

Minimize: $$z = \text{Max}\,(c_{ij}x_{ij})$$

where Max $(c_{ij}x_{ij})$ means the largest of all of the products $c_{ij}x_{ij}$. The following algorithm solves this problem effectively:

Step 1 Rank all the c_{ij}, with the smallest c_{ij} being given the rank 0, the next smallest being given the rank 1, and so forth; if two or more of the c_{ij} are equal, break the tie arbitrarily. Call these ranks r_{ij}, and note that no two r_{ij} are the same. Define $C_{ij} = 2^{r_{ij}}$. Set up an assignment problem in which each of the costs is given by C_{ij}. Solve this assignment problem. Stop.

As an example, we solve the problem of Figure 2-3, as shown in Figure 7-2. Because of the simple numbers in the problem, each of the ranks is just one less than the c_{ij}. You will note that the optimum solution, when transferred back to the original matrix, gives a z-value of 6; that is, the three times chosen have values of 1, 3, and 6; and the largest of these, namely, 6, is the slow one which will control the speed of the production line. This particular solution turns out to correspond to one of the two alternate optima that we found for the original problem, but the other one (assignment along the diagonal) is not an alternate-optimal solution to this problem, and in fact would have a z-value of 7.

Obviously this algorithm wouldn't work very well in practice on a large matrix. For example, in a 10×10 problem, the largest r_{ij} would be 99, and the largest C_{ij} would be 2^{99}, which is a 30-digit number. Better algorithms are available for this problem; see, for example, *Operations Research,* vol. 19, pp. 1747–1751.

c_{ij} matrix				r_{ij} matrix				C_{ij} matrix		
2	6	9		1	5	8		2	32	256
3	7	8	→	2	6	7	→	4	64	128
5	4	1		4	3	0		16	8	1

↓ −7

0	30	254
0	60	124
15	7	0

↓ −7

0	0*	231		0	23	254		0	23	254
0*	30	101	← −46	0	53	124	← +23	0	53	124
38	0	0*		38	23	23		15	0	0

FIGURE 7-2
Solving a bottleneck assignment problem.

It is of some interest to note why this algorithm works. It works because of the simple inequality

$$\sum_{i=0}^{n-1} 2^i < 2^n \qquad (7\text{-}9)$$

In other words, when we solve the final assignment problem, and attain a certain minimum sum, that sum all put together must be less than the next single largest cost in the matrix.

Problem

7.4.1-1 (*a*) Solve the following bottleneck assignment problem:

7	8	5
5	2	8
8	5	3

(*b*) Why is it necessary to break ties?
(*c*) Try this problem without breaking ties; that is, each of the 5s becomes $r_{ij} = 2$, and so forth.
(*d*) How then can you find alternate optima?

7.4.2 The Quadratic Assignment Problem

As we mentioned in the beginning of Chapter 3, real applications of the assignment model are very rare; in fact, Problems 2-2 and 2-6 are virtually the only real applications we have ever seen (Problem 2-3 is not very realistic). All other applications either have many identical rows and/or columns (that is, they are transportation problems), or they have special restrictions which turn them into traveling salesmen problems (Section 7.5)—or, they have various other special restrictions. We talk now about one of these restrictions in a real application.

In this case we are assigning jobs (together with their appropriate production facilities) to work areas in a factory. The costs in this case arise not from putting a particular job in a particular area but from the interaction between different jobs in different areas. Specifically, let there be n different jobs in n different areas. We will use the subscripts i and k for jobs, and the subscripts j and l for areas. Now if we assign job i to area j and job k to area l, there is a cost c_{ijkl} due to the amount of traffic which must flow between the two areas. This depends not only on the distance between the two areas but also on the amount of interaction between the two jobs (namely, the flow of material and personnel).

The mathematical formulation of this problem yields constraints identical to those of equations (2-1), but the objective function is

Minimize:
$$z = \sum_{i=1}^{n} \sum_{j=1}^{n} \sum_{k=1}^{n} \sum_{l=1}^{n} c_{ijkl} x_{ij} x_{kl} \tag{7-9}$$

There are, as before, exactly n^2 x's, of which n will turn out to be 1 and the remaining $n(n-1)$ will be 0. But there are n^4 c's, of which all c_{ijij} are 0 (because the cost of interaction of x_{ij} with itself is zero), and all other c_{ijkj} and c_{ijil} are set equal to M (because you cannot assign two jobs to the same area or two areas to the same job).

The problem is not "quadratic" in the sense in which we learned that term in high school, because each of the x_{ij}^2 terms has a coefficient of 0; but it is called the *quadratic assignment problem* because in the objective function each x is multiplied by another x, which gives "second-order" terms which are equivalent to quadratic terms. In any case the problem is nonlinear. It can be solved by certain modifications of the quadratic programming method mentioned in Section 7.7 or by a modification of the traveling salesman algorithm presented in Section 7.5. In the latter case, one first sets up a cost matrix with $n(n-1)/2$ rows (one for each pair of jobs) and $n(n-1)/2$ columns (for each pair of areas). Thus, if there are 10 jobs and 10 areas, the resulting traveling salesman matrix is 45×45. The details can be found in more advanced texts.

Problem

7.4.2-1 (Adapted from an article by Dennis R. Heffley.) In Problem 2-2 we showed that the selection of a swimming relay team might be modeled as an assignment problem. In extending this to a track relay team, we note that some runners are particularly good at leaving the blocks (that is, being first), while others are good at putting on a burst of speed to break the tape (that is, being last). In addition, baton passing is critical, and one runner may be particularly good at giving the baton to a second but poor at giving it to a third. Show how this situation is modeled as a quadratic assignment problem. Assume, as before, eight candidates and four legs in the race.

7.4.3 The Capacitated Transportation Problem

Very few variables in linear programming problems are really unrestricted (an exception is found in the dual variables of assignment and transportation problems which, as noted in Section 6.6, are not meaningful when taken individually but only when taken in pairs). Most decision variables are constrained to be nonnegative; if they are constrained to be greater than some negative lower bound, we shift them for purposes of the simplex algorithm so that they are constrained to be nonnegative, as we noted when discussing subroutine A in Section 4.5; for example, if $x_j \geq -10$, we substitute a new variable which is 10 greater than the old one. Similarly, if we had $x_j \geq 10$, we could add that as an extra constraint; but it would be easier to define a new variable, 10 smaller than the old, which is constrained only by nonnegativity.

In many real problems there are also upper limits on the values of certain variables. In the conventional form of a linear programming problem this can be handled by simply adding constraints of the form $x_j \leq u_j$, where u_j is the upper bound. However, as we have noted, additional constraints add considerably to the time required to solve a problem (much more so than additional variables), and certain special-purpose algorithms are available in this case which are easier. We shall illustrate this special-purpose algorithm by a modification of the primal algorithm for the transportation problem, where upper bounded variables are quite common; for example, we may have a constraint that $x_{27} \leq 13$ because the railroad (or pipeline, or whatever) connecting origin 2 to destination 7 is limited in its capacity.

We note first that in this case we are not even guaranteed that there will always exist a feasible solution. Furthermore, if the capacity constraints are exceedingly tight (so that a feasible solution barely exists), it may not be easy to find a first feasible solution, or even to be sure whether or not one exists. Clearly the techniques of Section 7.1 are applicable; there are other, easier methods which we shall not go into. We assume that the upper bounds have been entered into the appropriate cells of the matrix and that a first feasible solution has been found, such that $0 \leq x_{ij} \leq u_{ij}$ for each i and j, where x_{ij} is the quota and u_{ij} is the upper bound. The algorithm proceeds almost as before. There must be exactly $n + m - 1$ basic variables. Now a basic variable is defined to be one for which the quota is $0 < x_{ij} < u_{ij}$; before, we required only that $0 < x_{ij}$—that is, that x_{ij} be strictly positive. As before, epsilons may be added if the solution is degenerate; now we can change either $x_{ij} = 0$ to $x_{ij} = \epsilon$ or $x_{ij} = u_{ij}$ to $x_{ij} = u_{ij} - \epsilon$. As before, for each nonbasic variable, the unique closed chain must be found in which all other variables are basic, and this chain (or the modi method) is used to determine an opportunity cost c_j. In going through the chain, the basic variable which becomes nonbasic is either the first one to become 0 (as before) or the first one which reaches its upper bound, whichever comes first. If both happen at once, we must add an epsilon; and if two variables reach their upper bounds at once, one of them must be given a quota $u_{ij} - \epsilon$. Finally, in looking for that nonbasic variable which should become the incoming variable, we looked before only for variables whose quota was 0 and whose opportunity cost was negative; now we must also look for a nonbasic variable whose quota is equal to its upper bound and whose opportunity cost is positive, in which case we will reduce its quota in going through the closed chain. The test for optimality is also modified in an obvious way; namely, the solution is optimal if the c_j is nonnegative for each nonbasic variable whose quota is 0 and nonpositive for each nonbasic variable whose quota is equal to its upper bound. The algorithm is really quite easy in practice; the solution is still guaranteed to be in integers if the upper bounds are all integers (and the other previous requirements are also met).

A similar modification of the simplex algorithm is available to handle upper bounded variables.

Problems

7.4.3-1 Solve Problem 3-1 with the upper bounds $u_{31} = u_{12} = u_{23} = 3$; all other $u_{ij} = M$ (that is, ignore them). You will have to work out some formal method of keeping track of all the data. For example, you might try writing the c_{ij} in the center of each cell, the u_{ij} in the lower left corner of each cell, the x_{ij} in the upper left corner, the c_j in the lower right corner of each nonbasic cell, and N or B (for nonbasic or basic) in the upper right corner of each cell. Put c_{ij}, u_{ij}, all α_i and β_j (at left and above), and $u_1 = 0$ in ink; other u_i, all v_j, all x_{ij}, all c_j, and the N's and B's should be in pencil.

7.4.3-2 Write the algorithm for the capacitated transportation problem. Use steps 0 through 4 and 7 through 10 of the primal algorithm of Section 6.5, and steps 5, 6, and $6\frac{1}{2}$ of the modi method, showing only the necessary changes.

7.4.4 The Fixed-Charge Transportation Problem

In the fixed-charge transportation problem, in addition to the previous cost $\Sigma\Sigma c_{ij}x_{ij}$, there is an additional charge δ_{ij}, which is incurred if the route ij is used and not otherwise. That is, there is an additional fixed charge δ_{ij} if and only if $x_{ij} > 0$. (There is a corresponding fixed-charge linear programming problem in which there is a fixed charge δ_j if and only if x_j is greater than 0.) Fixed-charge problems, which are nonlinear, are unfortunately very common and very difficult. You can see how such a problem might arise. There could be a fixed charge for using a railroad car, after which the cost is linear with the amount of material loaded in the car. The fixed charge is even clearer in Problem 3-3, where we were transporting requirements from one time period to another; here we can avoid setup costs by avoiding production of a particular item in a particular month.

Technically such problems are difficult because the corresponding cost function is "concave," which roughly means that if you graph the total cost against the amount, the function is concave when viewed from below, as defined more precisely in Section 7.7.1. We can more easily see in terms of the bedbug analogy why these problems are difficult. In all linear programming problems, if the bedbug were at a vertex (a basic solution) which was not optimal, he could always find an edge along which to move in such a way as to reduce his cost; that is, he could start moving through nonbasic solutions (along edges) while continously reducing the cost. If there were no direction in which he could go and immediately reduce the cost, then he would be at the optimum. However, in this problem, as soon as he leaves a vertex, he must immediately incur a new fixed charge, which increases his cost; and there is no easy way for him to tell (remember, he's nearsighted) whether, by the time he gets to the next basic solution, he will have made up for it. It is still true that the optimal feasible solution will be at a vertex (will be basic). In the special case where all the fixed charges are equal, the solution to the problem without fixed charges would be the same as the solution to the problem with fixed charges, provided the optimum solution is not degenerate; however, since a degenerate solution would save on one or more fixed charges, this introduces an extraordinary complexity. In the case where the various charges are different, the problem is extremely difficult.

7.4.5 The Transshipment Problem

Thus far in examining transportation problems, we have assumed that the origins and destinations are quite separate. But it may happen that the most convenient way of shipping from a particular origin to a certain destination is to ship first to another origin, which then becomes a transshipment point. More generally, the transshipment problem arises when one is shipping from factories to warehouses to retail outlets, in which case each warehouse is a transshipment point. If one is considering different locations for warehouses, such a transshipment model must be analyzed for each possible set of warehouse locations so that the load of computation becomes very heavy indeed unless we have an efficient algorithm. Finally, in some cases we may be required to choose an optimal path through a whole network of transshipment points.

We do indeed have several effiicient algorithms for this problem. We use the following terminology: If there is a net amount to be shipped from a particular location (that is, if there is more to be shipped from it than to it), it is called an *origin*, whether it is a transshipment point or a source; if there is a net amount to be shipped to a particular location, it is called a *destination*. An origin which cannot act as a transshipment point is called a *source*, and a destination which cannot serve as a transshipment point is called a *sink*; all locations other than sources and sinks, whether they be origins, or destinations, or merely points which ship in as much as they ship out, are called *transshipment points*.

One method of solution which is conceptually very simple is to examine each possible pair of origin-and-destination and to determine the cheapest cost of shipment between them. These numbers are then entered as the c_{ij}'s in an ordinary transportation problem, with one row for each origin and one column for each destination. However, if there are many origins and many destinations, and if the network is complicated (so that it is possible to ship from an origin to a destination via several transshipment points), the determination of these c_{ij}'s may be very tedious. In this case we set up a transportation problem in which there is a row for each transshipment point and each source and a column for each transshipment point and each sink. Enter the appropriate α_i for each origin and the appropriate β_j for each destination. Then for each transshipment point add to both its α_i and its β_j some fixed large number—for example, the total amount to be shipped from all origins. For each transshipment point, enter $c_{ii} = 0$; for all other routes, enter c_{ij} where it is appropriate, and otherwise enter M (in other words, M for those routes which are not permitted by the conditions of the problem). This is now an ordinary transportation problem which may be solved by any of the algorithms we have given previously. The final solution will contain numbers (perhaps large numbers) for most or all of the x_{ii}, which are meaningless. But the remaining x_{ij} represent the optimal amounts to be shipped from point i to point j. By adding the large (but equal) imaginary amounts to both discrepancies at each transshipment point, we convert the transshipment problem into a conventional transportation problem.

Problems

7.4.5-1 In the following problem we have three origins, represented by O_i, four destinations, represented by D_j, and four warehouses, W_k, each representing a transshipment point.

	α_i	k = 1	2	3	4
α_i	16	23	27	39	29
	7	32	31	22	23
	8	24	22	18	26

k	β_j = 6	5	9	11
1	23	34	27	23
2	31	18	26	34
3	26	24	34	28
4	19	20	34	25

(*a*) Solve this problem by the first algorithm suggested in the text; that is, set up a 3 × 4 matrix of costs c_{ij}, each c_{ij} representing the minimum cost for transporting from O_i to D_j (incidentally this matrix should indicate which W_k yields this minimum cost), and then solve as an ordinary transportation problem.

(*b*) Solve by the second algorithm suggested in the text so that this becomes a 7×8 transportation problem.

7.4.5-2 A cash-management problem. (From an article by V. Srinivasan in *Management Science*, vol. 20, pp. 1350–1363, 1974. Srinivasan asserts that the resulting 14×13 transportation problem was solved on a Univac 1108 computer in 0.1 second, 30 times as fast as solving the corresponding linear programming problem.)

A firm must decide how to manage its cash flow over the next 6 months. Since it intends to reoptimize each day, the 6-month planning horizon is divided into unequal time intervals of 1, 1, 10, 20, 60, and 90 days, respectively (the more distant time intervals being less critical). The firm holds securities of face value 10, 10, 120, 50, 100, 250, and 500 thousand dollars, maturing, respectively, in periods 1 through 7 (7 is beyond the planning horizon). There are stated penalties for selling each security before maturity; these vary irregularly, but you may assume each of them to be known for your model. Purchases are predicted to be 50, 30, 400, 600, 1500, and 5700 thousand dollars, respectively, in periods 1 through 6; they must be paid within 30 days, and 2% discount is allowed if they are paid within 10 days. Spare cash may be invested at 4.8% per annum. A line of credit up to 2.5 million dollars is available at 7.2% per annum; such money must be held for 1 year (no short-term borrowing is possible), but for this model you should consider only the interest to the end of the planning horizon. There are other sources and uses of cash (not subject to decision) amounting to net cash outflows of 300, 260, 620, and 1050 thousand dollars in periods 1 through 4, respectively, and net cash inflows of $5,900,000 and $6,550,000 in the last two periods. There is $100,000 on hand at the beginning of period 1 (including the $10,000 security maturing in the period), which barely meets the minimum requirement of $100,000 cash on hand at all times. Formulate as a transshipment problem. It is easiest to formulate directly as a transportation problem, using the second technique suggested in the text; thus cash in each period will be both an origin and a destination (and you should add $30 million to each α and β). A dummy column (sources not utilized) is necessary and a dummy row (with all M's) will be useful in postoptimality analysis.

7.4.6 The Generalized Transportation Problem

Suppose we have n products to be made ($j = 1, \ldots, n$), with β_j units of the jth product to be produced. Furthermore we have m machines ($i = 1, \ldots, m$) which can manufacture these products, with the capacity of the ith machine being α_i. Suppose that it costs c_{ij} dollars to produce one unit of product j on machine i. Thus far this is a routine transportation problem if the sum of the capacities happens to be exactly adequate to manufacture all the products; if there is excess capacity, it can be easily converted into a routine transportation problem by adding a column of slack variables. However, suppose further that the different machines have different efficiencies on the different jobs. Specifically, suppose that to produce one unit of product j on the ith machine requires e_{ij} units of time, where e_{ij} is an inverse measure of efficiency. Other things being equal, a change in e_{ij} will cause a corresponding change in c_{ij}, but in this problem other things are not equal and the e_{ij} and c_{ij} vary independently. Now the problem is different. The formulation (3-1) must be modified by changing the first constraint from $\Sigma x_{ij} = \alpha_i$ to

$$\sum_j e_{ij}x_{ij} \leq \alpha_i \qquad (7\text{-}10)$$

This formulation is known as the *generalized transportation problem*. The problem can be even further generalized by making a similar modification to the second set of constraints, with additional constants, but this single generalization complicates the problem quite enough for us.

It turns out that we can solve this problem by a modification of the modi algorithm of Section 6.7, but there are certain significant changes. In the first place, as in the capacitated transportation problem, we are not guaranteed that there will be a feasible solution; and if the capacities are barely adequate, it may not be easy to find a first feasible solution. In the second place, even if there is a feasible solution, in general the solutions will not come out in integers. If fractional solutions are not acceptable, this problem becomes extremely difficult. We shall assume that fractional solutions are acceptable (because, for example, the numbers x_{ij} will turn out to be very large, in which case we may round to the nearest integer, or because the numbers x_{ij} are measured in something like gallons, which really are divisible).

We first note that this problem is not degenerate, because equation (3-2), which is the same as equation (6-3), does not hold. Thus there will be $n + m$ basic variables in a basic feasible solution. Next we note that because the constraints (7-10) are in the "wrong direction," each of the row dual variables is constrained to be nonpositive. That is,

$$u_i \leq 0 \qquad i = 1, \ldots, m$$

The variables v_j remain unconstrained. Finally there will be no arbitrary dual variables in this case. In the modi method we got started by setting $u_1 = 0$; in this case we will get started by setting $u_i = 0$ in each row where $s_i > 0$, because, as we already know, when a constraint is inactive the corresponding dual variable must be equal to 0.

Thus we set up a transportation matrix just as before in the modi method, adding an $(n + 1)$st column for the slack variables; this $(n + 1)$st column does not have a β_j, and the sum of the x's in this column is meaningless. In each cell we must enter the appropriate e_{ij} as well as the c_{ij} (and, eventually, the x_{ij} for basic variables and c_j for nonbasic variables); in column $n + 1$, each $c_{ij} = 0$ and each $e_{ij} = 1$. We then make a first feasible assignment of quotas by the northwest-corner rule or any other appropriate method (we are assuming that a feasible solution can be found), remembering that the row feasibility must be determined by equation (7-10). Next we set $u_i = 0$ for each row where $s_i > 0$. And finally we compute all the v's and the remaining u's by an obvious modification of the same formula we used before, namely,

$$c_{ij} - e_{ij}u_i - v_j = 0$$

for each basic variable. As before, there will be exactly enough basic variables to determine each of the u_i and each of the v_j uniquely. Note that there are just $n + m$ basic variables and the same number of u's and v's to be determined (no v is computed for column $n + 1$); each basic variable in column $n + 1$ determines a $u_i = 0$, and each other basic variable determines one u or one v.

We compute the opportunity costs for each nonbasic structural variable as before, namely,

$$c_j = c_{ij} - e_{ij}u_i - v_j$$

where, of course, all the u_i are nonpositive, while the v_j are usually (but not always) positive. As before, the most negative c_j becomes the incoming variable. At this point, however, things get very messy, because we cannot always determine a chain as before; in fact, there may exist closed chains of basic variables in this problem. To determine the outgoing variable, we must do something equivalent to comparing ratios b_i/a_{ij} in the simplex algorithm. This computation is straightforward but tedious, so we do not pursue it. We merely wished to indicate that the elegant properties of the transportation model and its associated algorithms are fragile and may be destroyed by seemingly minor changes in the problem.

While this model has historically been called the generalized transportation problem, that name could equally well have been given to most of the other models which were discussed in this section, and to dozens of other models which we have omitted.

7.4.7. Network Models and Graph Theory

We have noted that the transshipment model might involve a considerable network of transshipment points; in that case a special trick was required to convert the problem into a conventional transportation problem. It turns out that there are a wide variety of important network models in operations research and that many of them can be converted into assignment, transportation, or related models, and vice versa. The transportation problem itself, and more especially the capacitated transportation problem, can be considered a network problem in which each origin is a source, each destination is a sink, and the network consists of pipelines between the sinks and sources with some or all of these pipelines having finite capacities. The traveling salesman problem of Section 7.5 is a kind of network problem in which the salesman must find a shortest path (through the network) which visits each node and returns to the starting node.

One of the most important of all network problems is that of finding the shortest path through a network; for example, given a network (which might be a few cities as in the traveling salesman problem, certain cities being connected by paths characterized by a particular length or cost), the problem is to find the shortest (or lowest cost) path through the network from a particular designated origin to a particular designated destination. As we shall see, the traveling salesman problem is moderately difficult; but without the constraint of the salesman having to visit each city, the shortest route problem is pretty easy. It can clearly be modeled as a transshipment problem, with a single unit available at the single source and required at the single destination; and when this is converted into a conventional transportation problem by the algorithm of Section 7.4.5, each of the resulting origin or destination requirements will be one, and thus this will be an assignment problem. Conversely, assignment problems can be modeled as shortest route problems, and there are network algorithms for solving them. It turns out that this shortest route model has important applications, for example, in scheduling problems.

Not every linear programming problem can be modeled as a network problem; conversely there are a wide variety of network problems which cannot be modeled as linear programming problems, or for which there are easier solutions in other ways. Among the best known of these is finding the longest path through a network, which turns out to be the key step in the models called PERT (Program Evaluation and Review Technique) or CPM (Critical Path Method).

Another extremely important application of network theory is to electric circuits, which consist of networks including energy sources, energy dissipating elements (resistances), and energy storing elements (inductances and capacitances). This is an enormously complicated subject in its own right, the subject of numerous textbooks which are longer and more difficult than this. There is a branch of mathematics called "graph theory," or "the theory of linear graphs," which is specifically applicable to networks. It is rather formidable and makes heavy use of matrix algebra. Suffice it to say that this theory is underlined by the theory of duality. In particular, every graph (or network) which can be diagrammed straightforwardly has a dual in which each vertex (or node) of the primal becomes a loop of the dual, and vice versa; and it will not be surprising that there is a close relationship between this kind of duality and the duality that we have discussed arising from interchanging rows and columns of matrices. We cannot go deeply into this subject, but we can at least indicate that every network can be converted into a matrix and, therefore, that most network problems can be converted into a problem of the kind that we have become familiar with in this book. Specifically, if we have any network, consisting of vertices and paths between these vertices, then we can construct a matrix in which each row corresponds to a vertex and each column to a path. If the path is connected at the vertex, we write 1; otherwise we write 0. Such a matrix completely describes the geometry of the graph; obvious modifications can be made if the paths are unidirectional, or have certain capacities, and so forth. Conversely it is possible to convert any mathematical programming problem whose constraint matrix shows certain properties into a network model.

In Chapters 2 through 6 of this book we introduced the assignment, transportation, and general linear programming problems, analyzed them, and showed some of their relationships. In this section we have been indicating a relationship of these problems to other problems, and in this subsection in particular we are indicating a close relationship between these problems and network problems. We cannot pursue this subject any farther, except to point out that these kinds of relationships make mathematical modeling interesting, useful, and less than hopelessly difficult: As one learns more and more models, one meets old familiar ones ever more frequently so that it is possible to have a working relationship with most of the important ones. Thus, if you now undertook the study of network theory, you would find that you already know quite a lot of it from linear programming theory.

7.5 THE TRAVELING SALESMAN PROBLEM

The traveling salesman problem was posed by the American mathematician Hassler Whitney in 1934. Whether it has yet been solved depends on exactly what one means by "solve."

We are given an $n \times n$ matrix of c_{ij}'s in which each c_{ij} represents the distance from city i to city j, as in Figure 7-3. The salesman must start from the first city (Cleveland), visit each of the other cities once, and return to the first city, and he is to minimize the distance of his tour. That is, he must pick one element from the first row of the matrix, which represents going from Cleveland to some other city; then if he has selected the element $(1,j)$ he must pick one element from the jth row, corresponding to the next city which he visits after city j; and so forth. Finally, therefore, he must select n elements out of this matrix, exactly one from each row and one from each column, and he must minimize the sum of the n elements selected. So far the problem looks exactly like the assignment problem; there are, however, some subtle differences. In the first place, the salesman is clearly not allowed to select any of the elements on the diagonal, which represent going from city i to city i. To prevent this, we enter the element M at each point along this diagonal, where M (as usual) represents a very large number. However, note another distinction: In the assignment problem there's nothing to prevent us from assigning the first worker to the second job and the second worker to the first job; but if the salesman goes from the first city to the second city, and then from the second city to the first, he never gets to the third city. Furthermore, if he goes from the first city to the second and from the second city to the third, he may not then return to the first city because then he never gets to the fourth city. Thus, in some sense, cyclic solutions are not permitted.

This prohibition on cyclic solutions is easy to verbalize but hard to formalize mathematically. There exists a linear formulation of the problem, but the number of linear constraints which must be written is forbiddingly large (of

	Cleveland	Cincinnati	Columbus	Pittsburgh	Detroit	Fort Wayne	Indianapolis
Cleveland	*M*	244	139	134	167	212	309
Cincinnati	244	*M*	107	284	265	154	108
Columbus	139	107	*M*	191	202	152	172
Pittsburgh	134	284	191	*M*	285	330	354
Detroit	167	265	202	285	*M*	160	277
Fort Wayne	212	154	152	330	160	*M*	121
Indianapolis	309	108	172	354	277	121	*M*

FIGURE 7-3
Distances between certain American cities.

the order of $n!$). There exist, also, a number of integer programming formulations (Section 7.6). We shall not produce here any of the mathematical formulations (some are investigated in Problems 7.5-3 and 7.5-4) but simply present an algorithm for solving the problem.

This problem is by no means unimportant; not only does it have some rather obvious applications in terms of routing delivery trucks or salesmen, but it turns out to be exactly equivalent to the so-called job-shop sequencing problem. Suppose we have 20 jobs which must be done on a machine (say, a turret lathe), and the length of time which is required to set up the machine depends not only on the next job but also on the previous one. (The classical example: It's easier to clean the ice-cream machine after making vanilla ice cream and before making chocolate than vice versa.) Specifically in this case the number c_{ij} is the length of time required (or the cost in dollars, perhaps) for the setup if the previous job was job i and the next job is job j. Numerous job shops are faced frequently with just this type of problem, and the present state-of-the-art is such that the best method of sequencing in terms of cost effectiveness is generally to say to the foreman, "Joe, sequence these in the best order you think possible." The problem of minimizing the amount of wire required to connect various terminals in an electric circuit may be modeled as a traveling salesman problem; we have indicated that the quadratic assignment problem can be so modeled; and there are other applications.

For some reason this problem has become a classic. Almost every applied mathematician interested in operations research has tried his hand at solving it. A large number of algorithms have been developed, and there are a remarkable number of very ingenious translations of this problem into other problems which may be easier to solve. An indication of the complexity of the problem may be obtained from the following pseudotheorem: If there is a pair

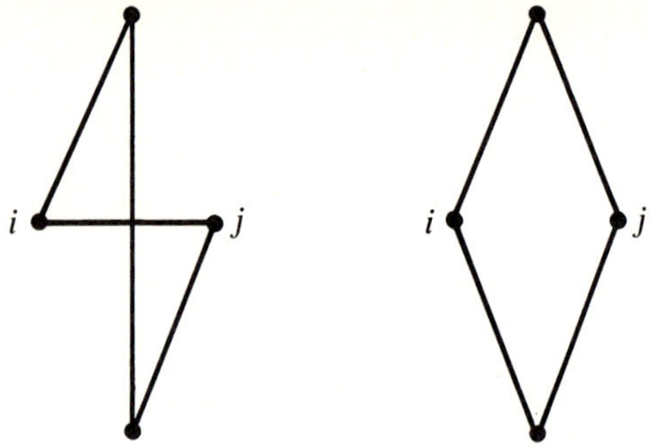

FIGURE 7-4
A four-city traveling salesman problem.

of cities, i and j, such that i is closer to j than any other city, and j is closer to i than any other city, then the direct connection between i and j will be part of the optimal tour. This conjecture is one of those obvious things which turns out to be false. The demonstration of its falsity can be seen from the four-city problem of Figure 7-4. Because of symmetry there are only two possible tours, and the one on the left which does include the direct link between i and j is clearly longer than the one on the right which does not.

7.5.1 Branch-and-Bound Algorithm for the Traveling Salesman Problem

First we present the algorithm; next we explain it and give an example; finally we assert that this kind of algorithm is applicable to a wide variety of problems.

Step 1 Set $z = M$. Subtract from each row its smallest element; subtract from each column its smallest element; keep track of the amounts subtracted; enter their sum at the first node in the master diagram. Go to step 2 or 3 as you prefer.

Step 2 Solve the resulting assignment problem by the dual algorithm. Enter the amount subtracted in the master diagram (by increasing the cost of the appropriate node). Go to step 3.

Step 3 For each 0 in the resulting matrix, compute and enter its *penalty*, which is the sum of the smallest other element in its column and the smallest other element in its row. Select the largest of these penalties; call its row i, its column j, and the penalty itself $P(ij)$. Go to step 4.

Step 4 Draw two branches in the master diagram. The branch on the right is called ij, the branch on the left is called $\overline{ij}$. For the left branch, enter the cost of the preceding node plus $P(ij)$. Determine the ineligible tour by subroutine A. Go to step 5.

Step 5 Delete row i and column j; set to M the one element indicated by subroutine A; subtract from every row its smallest element, subtract from every column its smallest element; enter in the most recent right-hand node on the master diagram the sum of the cost of the preceding node and the total amount just subtracted. Go to step 6.

Step 6 If the resulting matrix is 3×3 or larger, delete all penalties and go to step 2 or 3 as you prefer. If the resulting matrix is 2×2 go to step 7.

Step 7 Add the appropriate two pairs of nodes to complete the master diagram (the $\overline{ij}$ nodes will have cost M). If the cost of the final node is less than z, replace z by this number. Delete all nodes in the master diagram whose cost is greater than z (greater than or equal to z if you are not interested in alternate optima). Go to step 8.

Step 8 If there are no nodes left on the master diagram, the solution is complete; stop. Otherwise go to the highest node (or, if you prefer, to the lowest-cost node) on the master diagram, reconstruct the corresponding matrix by subroutine B, and go to step 2 or 3 as you prefer.

Subroutine A

In order to prevent a cyclic tour, exactly one element must be deleted from the matrix at this point (that is, changed to a value M). Find the longest subtour which is already present and which includes the element (i,j), and then delete (that is, change to M) that element which would allow this tour to become cyclic. If the element (j,i) is still in the matrix, then this is the element that must be changed to M. If it is no longer in the matrix (that is, if row j or column i or both have already been deleted), it is because the element (i,j) which has just been admitted to the tour is part of some longer subtour. For example, if the element (j,k) is already in the tour, and no element (k,l) or (l,i) is in the tour, then the element (k,i) is the one that must be changed to M to prevent a cyclic tour; if the elements (j,k) and (k,l) are already in the tour, but no element beginning with l or ending with i, then it is the element (l,i) which must be deleted.

Subroutine B

Go back to the original matrix and then modify it, going down the master diagram node by node to the chosen node and at each node carrying out the indicated operation on the matrix. If you are taking the left-hand branch ($\overline{ij}$), delete the zero in (i,j) from the matrix (set it to M) and subtract the smallest remaining element from row i and column j; in this case you do not have to go through subroutine A, since you have already set (i,j) to M. If you are taking a right-hand branch (ij), delete row i and column j, set the appropriate element to M by subroutine A, and then subtract from every row and column its smallest element. Solve the assignment problem each time if you did it before. While carrying out these steps you do not need to change the cost on the master diagram, because you have already done that. Now return to step 3.

There are obviously a lot of options in this algorithm; the idea (as in the dual assignment algorithm) is to subtract as much as possible. You may increase the amount subtracted in any step by going through the assignment algorithm if you prefer. It might be a good idea to go through step 2 the first time, and after that not bother with the assignment algorithm; or it might be best to use step 2. In either case, you should keep a copy of the matrix you used the first time you went to step 3 for subsequent uses in subroutine B. You also have the option of picking the node with the lowest cost in the diagram at the end of step 6, but we think it is easier if, as suggested, you continue down the right-hand branch of the master diagram.

Now let us explain what the algorithm is all about. It is called a *branch-and-bound algorithm.* It belongs to a class of algorithms which are called *implicit enumeration* algorithms. The idea is that there are an enormous number of feasible solutions to the problem—in fact $(n - 1)!$—far too many to enumerate explicitly. What we do is to make it unnecessary to enumerate most of them explicitly. Specifically, each time we get to step 4 we divide all the remaining possible solutions into two groups (this is the branching part of the algorithm), namely, those which do include the node (i,j) and those which do not include this node. Clearly these two groups are mutually exclusive and collectively exhaustive. We then place a lower bound on the cost of every tour belonging to each group (this is the bounding part of the algorithm); that is, we compute a number (which we enter on the appropriate node of the master diagram) which is as big as we can make it, and still assert that every tour in the group must cost at least this much. Eventually we find a solution whose cost is z (in step 7), and then any group of solutions whose lower bound is higher than z has been implicitly enumerated and need no longer be considered.

The following example illustrates how the algorithm works. We will work with the problem of Figure 7-3, which shows real distances taken from an atlas. We will skip step 2 altogether. The results of applying step 1 are shown at the top of Figure 7-5; normally we would make a copy of this matrix so that we could mark it up in going through the algorithm and still have it available when we need it again in subroutine B. In this simple problem, however, we shall not have to go back to it. The total amounts subtracted from rows and columns is 904, and we have entered this at the top of the master diagram, which is commenced on the right of the matrix.

There is, of course, at least one 0 in each row and each column of this 7×7 matrix, and in this case there happen to be eight 0s. The penalty for each 0 has been computed and entered in Figure 7-5; we note that the largest of these is for $(i,j) = (1,4)$ and that $P(14)$ equals 84. We now branch: We divide all possible solutions in the problem into those which do include the element

M	110	5	$^{84}0$	$^{6}0$	78	175
137	*M*	$^{6}0$	177	125	47	1
32	$^{32}0$	*M*	84	62	45	65
$^{64}0$	150	57	*M*	118	196	220
7	105	42	125	*M*	$^{24}0$	117
91	33	31	209	6	*M*	$^{7}0$
201	$^{13}0$	64	246	136	13	*M*

~~*M*~~	~~110~~	~~5~~	~~$^{84}0$~~	~~$^{6}0$~~	~~78~~	~~175~~
137	*M*	$^{6}0$	177	125	47	1
32	$^{32}0$	*M*	84	62	45	65
M	150	57	*M*	118	196	220
7	105	42	125	*M*	$^{20}0$	117
91	33	31	209	6	*M*	$^{7}0$
201	$^{13}0$	64	246	136	13	*M*

~~*M*~~	~~110~~	~~5~~	~~$^{84}0$~~	~~$^{6}0$~~	~~78~~	~~175~~
130	*M*	$^{1}0$	177	119	47	1
25	$^{25}0$	*M*	84	56	45	65
M	93	$^{55}0$	*M*	55	139	163
$^{25}0$	105	42	125	*M*	$^{13}0$	117
84	33	31	209	$^{55}0$	*M*	$^{1}0$
194	$^{13}0$	64	246	130	13	*M*

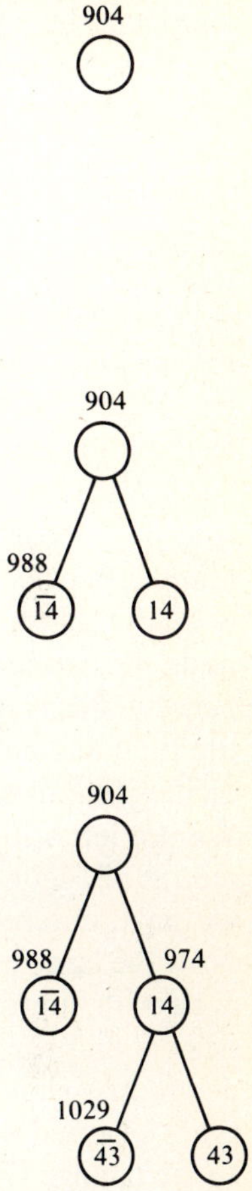

FIGURE 7-5
Solving the traveling salesman problem.

(1,4) and those which do not include this element. Clearly any tour which does not include this element will have to use some element from the fourth column, of which the cheapest is 84. Thus the total cost of any such tour must be at least $904 + 84 = 988$, which we have entered on the left branch of the master diagram. It is now possible that we will find some tour whose total cost is less than 988, in which case none of these tours ($\overline{14}$) will have to be explicitly enumerated. You can now see why we pick the largest penalty; this gives us the maximum possibility of eliminating the whole group of tours without considering them in detail.

We are now going to investigate the tours which do use the element (1,4). If we are going to use (1,4), we cannot use (4,1) without cycling, and indeed the algorithm tells us (through subroutine A) to set (4,1) to *M*. Furthermore, we cannot use any other element in row 1 or in column 4, and we delete them, as shown in the next matrix. At this point it is apparent that any tour which does use (1,4) is also going to cost more than 904, because we deleted some zeros. Specifically, we happened to delete 0s in (1,5) and in (4,1). We therefore—in the third matrix—subtract 7 from column 1, 6 from column 5, and 57 from row 4, for a total of 70, and enter $904 + 70 = 974$ on the right branch of the master diagram. We also recompute on this matrix all penalties. The largest penalty at this time is 55; since there are two of these, we break the tie by the northwest-corner rule and pick (4,3) to branch on. We enter the appropriate two nodes on the master diagram, and on the left branch we enter $974 + 55 = 1029$. At this point subroutine A tells us that since we have elements (1,4) and (4,3) already in the tour, the element to be deleted is (3,1).

We now move to Figure 7-6. We have to change (3,1) to M and delete row 4 and column 3. On the figure we identify the rows and columns by number (we need the original row and column numbers for subroutine A). You will note that we have subtracted 1 from row 2 and entered $974 + 1 = 975$ on node 43 of the master diagram. We also recompute penalties. The largest penalty, 84, is in element (5,1), so we extend the master diagram as shown on the right and enter $975 + 84 = 1059$ on the left branch. Subroutine A now indicates that element (3,5) must be deleted; we also delete row 5 and column 1. We are left with only four rows and columns which we present in the second matrix of Figure 7-6. Note that we have subtracted 13 from column 6, and so we have entered $975 + 13 = 988$ over the node (51) on the master diagram. Now the largest penalty is 118 in (6,5); we change element (3,6) to M in accordance with subroutine A, and delete row 6 and column 5. Recomputing penalties we find the largest penalty on element (2,7), and so we extend the master diagram as shown. Now, since neither city 2 nor city 7 has been involved in the tours selected thus far, the element to be deleted by subroutine A is (7,2), and so we change (7,2) to M, delete row 2 and column 7, and find that we have arrived at step 7.

You may have noted that each time we go through the beginning of step 5 we delete two M's and add one M, and since we also make the number of rows and columns smaller by one, there is at each stage of the algorithm the same number of M's as there are rows (or columns); in fact, it turns out that there is always exactly one M in each row and exactly one M in each column (until you have gone through an ij node in subroutine B, after which there may be additional M's). Thus, when we reach step 7 as we now have, there will be two M's and two 0s, and we must pick the two remaining zeros (in either order) to complete the master diagram. We complete the master diagram, and we see that we have chosen the tour indicated below it and that its cost is 988. Checking back in Figure 7-3 to add up the cost of these elements (Cleveland to Pittsburgh to Columbus to Cincinnati to Indianapolis to Fort Wayne to Detroit to Cleveland), we see that the cost is indeed 988 for this tour.

This brings us to step 8, and we note that all remaining nodes on the matrix have cost higher than $z = 988$ with the exception of the node $(\overline{14})$, which is

	1	2	5	6	7
2	129	M	118	46	$^{46}0$
3	M	$^{45}0$	56	45	65
5	$^{84}0$	105	M	$^{13}0$	117
6	84	13	$^{56}0$	M	$^{0}0$
7	194	$^{13}0$	130	13	M

↓

	2	5	6	7
2	M	118	33	$^{33}0$
3	$^{32}0$	M	32	65
6	33	$^{118}0$	M	$^{0}0$
7	$^{0}0$	130	$^{32}0$	M

↓

	2	6	7
2	M	33	$^{98}0$
3	$^{65}0$	M	65
7	$^{0}0$	$^{33}0$	M

↓

	2	6
3	0	M
7	M	0

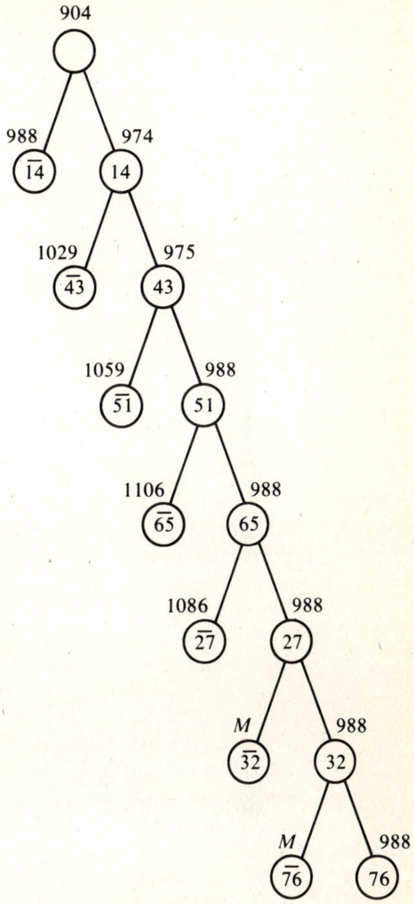

1 – 4 – 3 – 2 – 7 – 6 – 5 – 1

FIGURE 7-6
The completed solution.

equal to 988 (most problems, of course, would not be so easy). At this point, if we were interested in alternate-optimal solutions, we might have to go back to the top of Figure 7-5, set the element (1,4) to M, subtract 84 from column 4, and then return to step 3. However, in this case it is obvious what is happening: The alternate-optimal solution that would be developed if we did all that would be simply the same tour taken in the reverse order. This particular alternate-optimal solution arises because the original matrix was symmetrical (that is, every $c_{ij} = c_{ji}$), which would not happen in the job-shop sequencing problem but which will happen when the c_{ij}'s are distances on a plane surface.

The general concept of branch-and-bound algorithms, and the still more general concept of implicit enumeration, are techniques which are applicable to a wide variety of problems. We ask you in Problem 7.5-2 to attempt to design a branch-and-bound algorithm.

Problems

7.5-1 Solve Problem 2-1 as a traveling salesman problem. Use the c_{ij}'s given there, but set all $c_{ii} = M$. This problem will be a little more tedious than our example: The first trip down the right-hand branch of the master diagram happens in this case to yield a poor solution (z); hence, you will have many nodes left and must go through subroutine B several times. There are two optimal solutions; you will save time if you stop after finding one of them.

7.5-2 (*a*) Design a branch-and-bound algorithm for the modified transportation problem of Problem 6-14. This was a problem in which there were m origins (products to be stored), $i = 1, \ldots, m$, and n destinations (storage facilities), $j = 1, \ldots, n$, with the cost of storing the ith material in the jth facility being c_{ij} per unit. There is also a restriction that not more than one material may be stored in any one facility. In addition, there is an $(n + 1)$st facility in which any of the materials may be stored at an additional cost. The total capacity is more than adequate. We will add an $(m + 1)$st row to represent slack capacity.

The method of solving will be as follows. First set

$$\beta_{n+1} = \sum_{i=1}^{m} \alpha_i \qquad \alpha_{m+1} = \sum_{j=1}^{n} \beta_j \qquad \text{and all } c_{m+1,j} = 0$$

Next solve as a transportation problem without the nonlinear constraints. The cost of this solution gives a lower bound on the cost of the actual problem; this solution will probably be infeasible in the actual problem, in that there will be two or more positive x's in some column other than the $(n + 1)$st. Now pick some column, say, the jth, in which there are two or more positive x's. The feasible solution to the actual problem will have only one positive x in this column, and this positive x will be in the first row, or in the second row, or .... Branch on each of these possibilities (including those where $x_{ij} = 0$ in the solution just obtained). For the first branch, where x_{1j} may be positive and all other x_{ij} will be 0, leave c_{1j} alone and set all other c_{ij} to M; for the second branch set all c_{ij} to M except c_{2j}; and so forth. Solve each of these as an ordinary transportation problem. You will thus obtain m numerical values for the cost of these m different problems; the lowest of these supplies a lower bound for the entire problem, which becomes a new lower bound if it is larger than the previous lower bound. In addition, if one (or more) of these solutions turns out to be feasible in the actual problem, then its cost supplies an upper bound to the cost of the actual problem. All nodes whose cost is lower than the lower bound or higher than the upper bound can be eliminated. Now working with the problem corresponding to your present lower bound, pick another column with two or more positive x's, and branch again. Eventually a node will be found whose cost is the lower bound and the upper bound, which are then equal, and this will be the solution to the problem.

(*b*) Try your algorithm on the following problem:

		Storage space			
		30	70	40	M
	10	20	14	19	24
Storage demand	80	15	13	20	22
	70	18	18	15	22

Here $m = n = 3$. This problem is from *AIIE Transactions*, March 1971, pp. 37–43, where the solution algorithm is also presented.

7.5-3 (*a*) Formulate the traveling salesman problem mathematically in a form comparable to equations (2-1). You are given a matrix of c_{ij}'s in which $c_{ii} = M$; you are asked to find a set of x_{ijk}'s, where $x_{ijk} = 1$ if the salesman goes from the ith city to the jth city on the kth leg of this trip, and 0 otherwise. The trick, of course, is in writing the cyclicity constraints. This is done by ensuring that if leg k ends in city j, then leg $k + 1$ must begin in city j. To ensure this, a certain sum over x_{ijk} is set equal to a certain sum over $x_{jl,\ k+1}$ for certain values of j and k (not all values, because the trip must eventually end, and therefore cycle, after leg n).

(*b*) If you have written this in an obvious symmetrical form, you will have introduced some unnecessary variables and constraints and also introduced a large number of alternate-optimal solutions. For example, for any given tour that is optimal, you will have n different solutions starting that particular tour at any one of its n cities. To avoid this you should, for example, assume that the salesman leaves city 1 on the first leg of the tour, which means that you will wish to write something special for x_{1j1}. See if you can rewrite this to eliminate such possibilities.

7.5-4 Why does the following elegant formulation for the traveling salesman problem work? Take equations (2-1) and add:

Given (in addition to the c_{ij}'s): $c_{ii} = M \quad i = 1, \ldots, n$
Find (in addition to the x_{ij}'s): $u_i \quad i = 2, \ldots, n$
Subject to (in addition to the constraints already given):

$$u_i - u_j + nx_{ij} \leq n - 1 \qquad i = 2, \ldots, n;\ j = 2, \ldots, n;\ i \neq j$$

Hint: Consider a cycle such as $x_{23} = x_{37} = x_{72} = 1$. Why is it ruled out? Next, consider a cycle such as $x_{12} = x_{23} = x_{31} = 1$ (where $3 < n$); why is it ruled out? Then consider a complete tour (such as the solution to our example problem); why is it not ruled out? This last question must be answered by constructing a feasible set of u_i; it shouldn't be too hard to guess these u's if you have gotten the idea from the previous parts of the problem and realize that they must have something to do with the solution which we found above. These u_i will always turn out to be integers, whether or not they are so constrained, but the x's must be constrained to be 1 or 0.

7.6 INTEGER PROGRAMMING

We have seen a number of problems, including the assignment problem, the subsystem test problem, and the traveling salesman problem, in which the variables are constrained to be either 0 or 1. The assignment problem has the unusual property that we may ignore this constraint and still get the right answer, but, as we indicated in Section 6.4, this is not usually the case.

A wide variety of other problems can be modeled by introducing special variables which are restricted to be either 0 or 1. For example, the fixed-charge problem of Section 7.4.4 can be modeled by introducing a set of variables y_j, each of which can only be 0 or 1 (y_j will turn out to be 1 if x_j is positive and 0 if x_j is 0). Then we multiply each y_j by its associated fixed charge, and add these into the objective function. Finally, we add a set of simple linear constraints of the form $x_j \leq u_j y_j$, where u_j is an upper bound on the jth variable. These constraints ensure that each y_j will indeed be 1 if its corresponding x_j is positive, while the cost minimization ensures that y_j will indeed be 0 if x_j is 0.

There are many other comparable clever applications of "zero-one variables" to special problems. For example, we may have a problem in which either the first constraint must hold or the second constraint must hold, but not necessarily both. We saw an example of this in Problem 6-6. We already had (from Problem 3-6) $\Sigma a_{ij}x_j \leq b_i$, where b_i was the amount of the bid (it became the α_i of the transportation problem). We added the additional constraint that, for every i, either $\Sigma a_{ij}x_j \geq \frac{1}{2}b_i$ or $\Sigma a_{ij}x_j = 0$ must hold. This can be modeled in an otherwise linear fashion by introducing appropriate variables y_i which can take on only the values 0 or 1. Moreover, we add the constraints $b_i y_i + 2\Sigma a_{ij}x_j \geq b_i$ and $b_i y_i + \Sigma a_{ij}x_j \leq b_i$. If $y_i = 0$, the first constraint says that $\Sigma a_{ij}x_j \geq \frac{1}{2}b_i$ and the second is trivial; if $y_i = 1$, the first constraint says that $\Sigma a_{ij}x_j \geq 0$, while the second says $\Sigma a_{ij}x_j \leq 0$, which taken together require $\Sigma a_{ij}x_j = 0$ as desired.

In addition to such zero-one problems, there are many problems in which variables are restricted to take on integer values which may be 0, 1, 2, 3, ... For example, if x_1 represents the number of thousands of widgets and the answer comes out 1.73, that's perfectly acceptable, because that means we make 1,730 widgets; but if x_1 represents the number of widgets to be manufactured, we cannot make 1.73 widgets.

We first introduced the integer constraints in equations (2-1) in the form of a quadratic (nonlinear) equation. In most cases, however, it is more convenient simply to write the constraint in the form that "each x_j must be an integer," or that "each x_j must take on only the values 0 or 1," since these turn out to be easier to deal with than the quadratic formulation.

The difficulty with all such problems is that finding the solution to the associated linear problem and then rounding off to the nearest integer will not in general give the best integer answer. This is illustrated in Figure 7-7, which shows a two-variable two-constraint problem superimposed on the integer grid. The two constraints are indicated by the lines identified as $s_1 = 0$ and $s_2 = 0$, and the solution to the associated linear problem is their intersection. An objective-function line is graphed as a dotted line passing through this solution point. The best integer solution is seen to be at the point $x_1 = 1$, $x_2 = 3$, which yields a higher profit than would be obtained by rounding off to $x_1 = 4$, $x_2 = 2$ from the solution to the linear problem. In zero-one problems, the difficulty of "rounding off" to the nearest integer solution is even more obvious. If we have two zero-one variables, then the feasible solutions are at the four corners of the appropriate square; if there are three variables, the feasible solutions are at the eight corners of the cube; if there are n variables, the feasible solutions are at the 2^n corners of the hypercube. In general the solution to the associated linear problem will lie somewhere within the interior of the hypercube, and there is no obvious way of going to the best vertex.

A number of algorithms have been developed for solving integer programming problems. Some of these use branch-and-bound methods, similar to those discussed in Section 7.5. We will discuss here only one algorithm, developed by the American mathematician Ralph Gomory, which uses so-called cutting planes. It can be shown that this algorithm always works for any integer programming problem, and a minor modification of it works for problems in which some of the variables are required to be integers while others may take on fractional values. However, it is not necessarily the most efficient

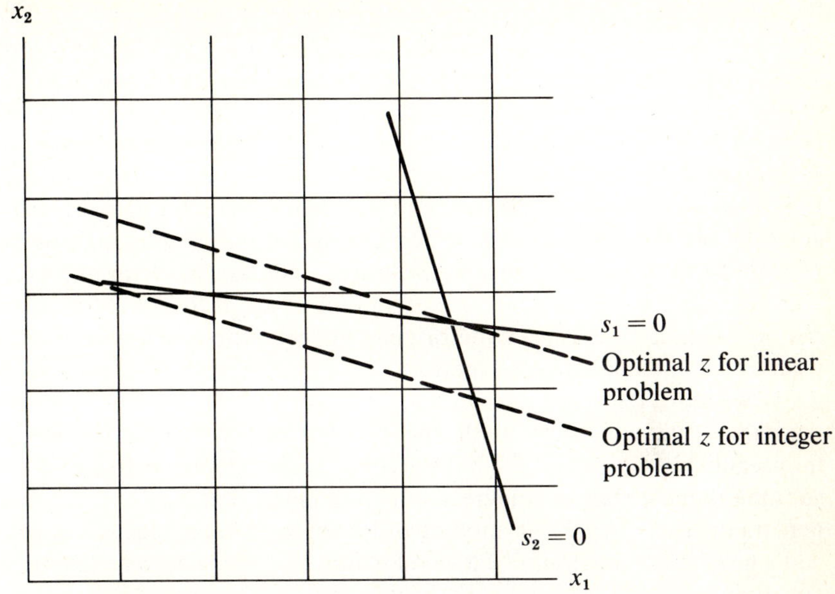

FIGURE 7-7
An integer programming problem whose optimal solution is far from the solution to the associated linear problem.

algorithm. If you're interested in solving real integer programming problems, you may have to learn a good deal more than is presented here.

7.6.1 The Gomory Cut

The idea of this algorithm is to solve a sequence of linear programming problems, each of which can be solved by one of the algorithms we already know, and each of which in some sense brings us closer to the final solution. More specifically, we start by ignoring all the integer constraints and solving the associated linear programming problem by any convenient algorithm. If the solution comes out in integers, we are finished; if not, we know that the true cost must be somewhat higher than the minimum cost we have just obtained. (Remember that we pointed out at the beginning of Section 4.6 that adding an additional constraint—in this case the integer constraint—can never help and might hurt.) We now add one new linear constraint, which must be obeyed by any integer solution and which is not obeyed by the present noninteger solution. This constraint removes part of the region of feasible solutions, including the present vertex, and presents a new linear problem, which may again be solved by a convenient algorithm to yield a new solution whose optimal cost is higher than before and in this sense is closer to the optimal feasible solution to the original integer problem. Again, if the solution to this linear programming problem comes out in integers, we are finished; if not, we add another linear constraint and repeat the procedure. It can be shown that this procedure converges in a finite number of steps to the optimal feasible solution to the original integer problem. However, in practice it may converge very slowly; that is, it may take a very large number of steps. Furthermore, while one gets closer and closer to the true optimal cost of the original problem, one does not necessarily get closer (in the sense of Figure 7-7) to the optimal feasible integer solution. Hence, if one is forced to stop before completion, one does not necessarily have a useful feasible solution. The fact that we are at all times working with a solution which is optimal (in terms of the local linear problem) but not feasible (in terms of the original integer problem) implies that we will wish to use the dual simplex method.

We will first solve an example problem and then formally present the algorithm; we will work with the problem (4-1), which we are already thoroughly familiar with. We have graphed this problem on Figure 4-8, and its solution is presented on Figure 4-11. Assuming that we have the additional constraint x_1, x_2 integer, the solution of Figure 4-11 is not feasible, and we must add a new linear constraint. To develop this constraint, let us examine the first constraint equation of our final tableau, which is

$$x_1 + \tfrac{2}{3}x_3 - \tfrac{1}{3}x_4 = \tfrac{5}{3}$$

We first break each coefficient into an integer part and a fractional part, where the fraction must be nonnegative and less than one. Note that the fractional part may equal 0 if necessary, but it must not equal 1. Thus, the coefficient of x_1, which is 1, breaks into an integer part which is 1 and a fractional part which is 0. The coefficient of x_3 breaks into an integer part which is 0 and a fractional part which is $\tfrac{2}{3}$; the coefficient of x_4 breaks into an integer part which is -1 and a fractional part which is $\tfrac{2}{3}$; and b_1 breaks into an integer part which is 1 and a fractional part which is $\tfrac{2}{3}$. Thus we rewrite the equation as

$$x_1 + \tfrac{2}{3}x_3 - x_4 + \tfrac{2}{3}x_4 = 1 + \tfrac{2}{3}$$

or

$$\tfrac{2}{3}x_3 + \tfrac{2}{3}x_4 = \tfrac{2}{3} + (1 - x_1 + x_4)$$

Now recall that this equation represents a particular vertex (when x_3 and x_4, the nonbasic variables, are 0); but it also represents a constraint which must be true for any solution to this problem even when x_3 and x_4 may become positive. It follows that for any solution to this problem the left-hand side of this equation must be nonnegative. Furthermore, for any feasible (integer) solution to this problem, the right-hand side must be equal to $\tfrac{2}{3}$ plus an integer. Finally, notice that this integer (the material in parentheses on the right-hand side) cannot be a negative integer, because then the entire right-hand side would be negative, which is forbidden. It follows that the material in parentheses must be either 0 or a positive integer. Finally, therefore, the left-hand side of this equation is equal either to $\frac{2}{3} + 0$, or $\frac{2}{3} + 1$, or $\frac{2}{3} + 2$, or . . . . Hence, for any integer solution to this problem,

$$\tfrac{2}{3}x_3 + \tfrac{2}{3}x_4 \geq \tfrac{2}{3} \qquad \text{or} \qquad -\tfrac{2}{3}x_3 - \tfrac{2}{3}x_4 \leq -\tfrac{2}{3}$$

This is the new constraint which will be added to the problem. Note that it does not hold for the present solution, where $\frac{2}{3}x_3 + \frac{2}{3}x_4 = 0$. Thus, we have introduced a cut (in this simple two-dimensional problem a cutting line, but in larger problems a "cutting plane"), which reduces the region of feasible solutions and which in particular removes the present vertex.

Appending this constraint, and adding the appropriate new slack variable x_5, we get

$$\begin{array}{cccccc} 1 & 0 & \frac{2}{3} & -\frac{1}{3} & 0 & \frac{5}{3} \\ 0 & 1 & -\frac{1}{3} & \frac{2}{3} & 0 & \frac{5}{3} \\ 0 & 0 & -\frac{2}{3} & -\frac{2}{3} & 1 & -\frac{2}{3} \\ 0 & 0 & \frac{1}{3} & \frac{4}{3} & 0 & -z + \frac{25}{3} \end{array}$$

which is in canonical form for the dual simplex algorithm. Iterating (on row 3 and column 3) we get

$$\begin{array}{cccccc} 0 & 0 & 1 & 1 & -\frac{3}{2} & 1 \\ 1 & 0 & 0 & -1 & 1 & 1 \\ 0 & 1 & 0 & 1 & -\frac{1}{2} & 2 \\ 0 & 0 & 0 & 1 & \frac{1}{2} & -z + 8 \end{array}$$

which is feasible.

You will note that each time we introduce a cut we add one row and one column to the matrix. Fortunately, the problem does not grow indefinitely. Whenever a new slack variable (any slack variable introduced with a Gomory cut) becomes basic, it may be dropped, as may the row stating its value. Thus, if we initiated the Gomory-cut procedure with m rows and n columns in the A-matrix, the A-matrix will never have more than $n + 1$ rows and $2n - m + 1$ columns. The reason why such a variable (and its row) may be eliminated when it becomes basic is similar to the reason why we were permitted to drop an artificial variable when it became nonbasic: Once such a constraint has been dominated by newer constraints, it can never again become active, and so it is superfluous.

In selecting the equation with which to work, if there are several fractional x's, it is desirable to select one which will "cut" as deeply as possible into the feasible region. While there is no entirely satisfactory way of doing this, the best rule seems to be to select that b_i with the largest fractional part.

Algorithm for All-Integer Linear Programming Problem

Step 1 Ignore the integer constraints. Solve the linear programming problem. Go to step 2.

Step 2 If every b_i is an integer, go to step 7. If one or more b_i are fractions, go to step 3.

Step 3 For each b_i, define Ib_i and fb_i so that $Ib_i + fb_i = b_i$, and Ib_i is an integer, and $0 \leq fb_i < 1$ (these definitions are unique). Choose that row for which fb_i is largest (break ties arbitrarily or by choosing the topmost row). Go to step 4.

Step 4 In this row, for each nonbasic j, define Ia_{ij} and fa_{ij} so that $a_{ij} = Ia_{ij} + fa_{ij}$, Ia_{ij} is an integer, and $0 \leq fa_{ij} < 1$ (these definitions are unique). Now write a new constraint $-\Sigma fa_{ij}x_j + s_k = -fb_i$, append this to the solved linear programming problem from step 2, fill in column k with zeros, and go to step 5.

Step 5 Solve by the dual simplex algorithm. If the incoming basic variable is a variable introduced in step 4, go to step 6. If the incoming basic variable was already in the problem in step 1, go to step 2.

Step 6 Delete the row and column of this basic variable. Go to step 2.

Step 7 The solution is complete. Check. Stop.

Problems

7.6-1 Add to Problem 4-9: X, Y, Z integers. Solve. Only one cut will be required.

7.6-2 (*a*) Maximize: $3x_1 + 4x_2$

Subject to:
$$x_1 + 2x_2 \leq 10$$
$$3x_1 + x_2 \leq 8$$

Three cuts will be required.

(*b*) Express each cut as an inequality in terms of the structural variables only (the slack variables may be eliminated by solving their defining equations for them and then substituting; for example, from the first constraint: $x_3 = s_1 = 10 - x_1 - 2x_2$). Graph the original problem and then each cut, to follow the progress of the problem.

7.6-3 Reexamine Problem 4-13. What do you think?

7.7 NONLINEAR PROGRAMMING

Nonlinear programming is, of course, any programming problem in which the objective function, or one or more of the constraints, or both are nonlinear functions of the decision variables; that is, what would otherwise be a linear programming problem except that not everything is linear. While integer constraints (Section 7.6) are nonlinear constraints, we consider them separately; that is, we talk about linear integer problems and nonlinear integer problems. The ones we discussed in Sections 7.5 and 7.6 are linear integer problems.

In several respects, nonlinear programming is enormously more difficult than linear programming. One obvious difference is that in linear programming, if there is an optimal solution, there must be an optimal solution at a vertex. In nonlinear programming the optimal solution may be anywhere on the boundary of the region of feasible solutions, or it may even be in the interior. More signif-

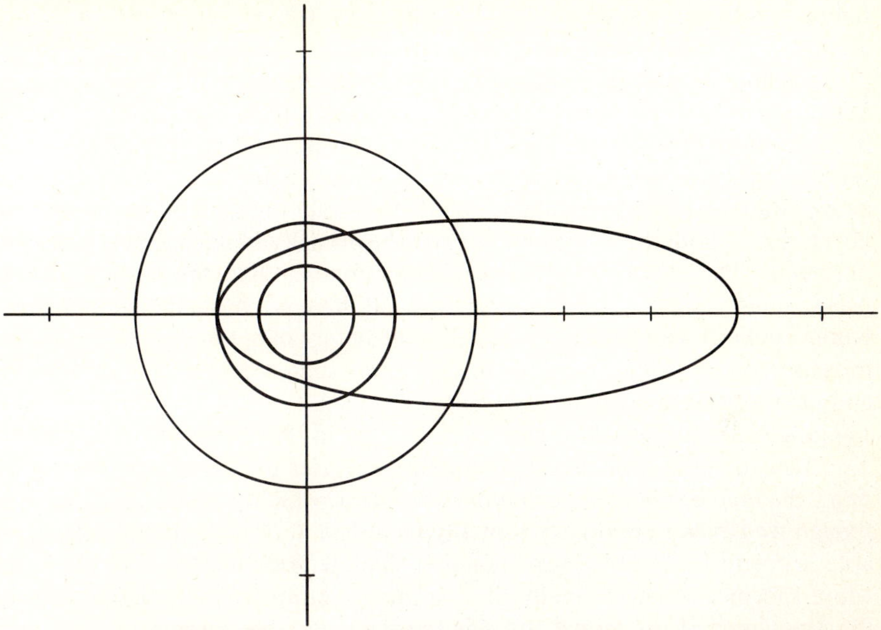

FIGURE 7-8
A constraint ellipse, $(x-2)^2/9 + y^2 \leq 1$, and a set of objective function circles, $z = x^2 + y^2$, for $z = \frac{1}{4}$, 1, and 4.

icantly and less obviously, in linear programming a "local optimum" was always an optimal solution to the original problem, but in nonlinear programming this is not necessarily true. We have already discussed this in Section 7.4.4. A *local optimum* is a point from which one cannot make any improvement by moving a very small distance in any direction. If you like, it is the top of a hill (in fact, there are certain optimization techniques useful in nonlinear programming that are called "hill-climbing" techniques). In linear programming we are assured that a local optimum is a *global optimum*; that is, if we have reached the top of the hill, we know that there is no valley across from which is a mountain that is even higher. In nonlinear programming this need not be true. Thus, the bedbug who finds himself at a point from which all possible directions are going up does not necessarily know that he has solved his problem. We can illustrate all this if we consider a constraint $(x-2)^2/9 + y^2 \leq 1$, which is the interior of the ellipse of Figure 7-8, and the objective functions $z = x^2 + y^2$, which are circles centered at the origin, three of which are also shown in Figure 7-8. If we are trying to maximize z, then clearly the west end of the ellipse (the point $x = -1, y = 0$) is a local optimum. The value of z at this point is $z = 1$, and no nearby feasible point has a z-value as large as this. However, the global optimum—that is, the true maximum for this problem—is at the east end of the ellipse ($x = 5, y = 0$), where $z = 25$. If we are trying to minimize z, then the optimum is at the origin ($x = 0$, $y = 0$), where $z = 0$, and this point is not on the boundary of the constraint set.

In nonlinear programming, as in linear programming, corresponding to any problem having n variables and m constraints, there is a dual problem having n constraints and m dual variables. However, there is a big difference. In the case of linear programming, the dual problem requires only the m dual variables. In nonlinear programming the dual problem in general has $m + n$ variables; that is, both the dual variables and the primal variables appear in the dual problem. The theory of nonlinear programming, like the theory of linear programming, is associated with duality, and there is a basic theorem called the Kuhn-Tucker Theorem which describes the nature of optimal solutions and permits us to develop algorithms for solving many such problems; but this theorem and the resulting algorithms involve advanced calculus (partial derivatives and Lagrangian multipliers) and cannot be treated in this book.

One theorem from linear programming carries over to nonlinear programming, the *theorem of complementary slackness*, which is really very simple, although we haven't previously formally identified it. It says that if a slack variable is positive, then the corresponding shadow price is zero, and vice versa. More specifically and formally, if x_j is the structural variable associated with the jth column of the primal and s_j is the slack variable associated with the jth row of the dual, then in any optimal basic solution $x_j s_j = 0$, which implies that at least one of these variables must be zero (if both are zero, there is a degeneracy); and of course, $y_i s_i = 0$. In the transportation problem of equations (3-1) and (3-3), the corresponding theorem is $x_{ij}(c_{ij} - u_i - v_j) = 0$, which says that quotas can be assigned only to zero costs.

Many of the algorithms of nonlinear programming result (as does the Gomory-cut algorithm) in setting up a sequence of linear programming problems to be solved by the simplex or a related algorithm. In some cases these algorithms can be shown to converge in a finite number of steps; in other cases the algorithms may go on forever, and one simply has to decide after a certain time what is "close enough." Whether a particular nonlinear programming problem is difficult or easy depends on the number of nonlinearities and their exact nature. For example, if all the constraints are linear, and the objective function is quadratic, then the problem (called a *quadratic programming problem*) is comparatively easy if the objective function has the proper kind of curvature. We now turn to a study of the kinds of curvatures that characterize nonlinear functions.

7.7.1 Convexity and Concavity

An important property of all linear programming problems, which we have hinted at but not previously stated formally, is that the region of feasible solutions is convex and, more specifically, is a convex polygon. In Figure 7-9*a* we show a convex polygon and in Figure 7-9*b* a nonconvex polygon. This concept was introduced in high school, where we defined a convex polygon as one that

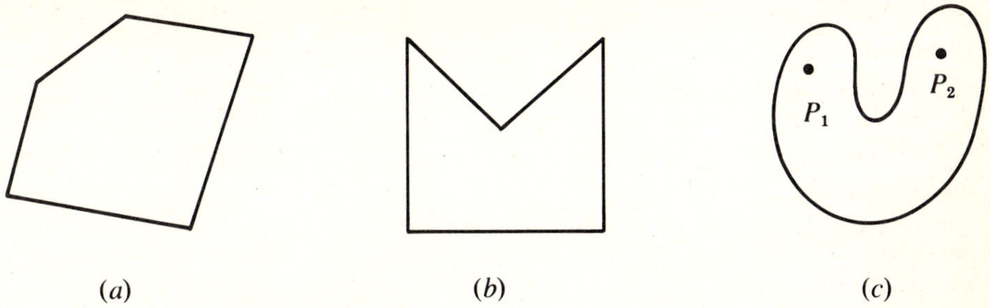

FIGURE 7-9
Convexity and nonconvexity. (*a*) Convex polygon; (*b*) nonconvex polygon; (*c*) nonconvex figure.

had no interior angle greater than 180°. However, such a definition is difficult to apply to many-dimensional polygons and breaks down completely if some of the boundaries are curved as in Figure 7-9*c*. Furthermore, we should like a definition of convexity which can be applied algebraically as well as geometrically. We therefore define a *convex set* as follows: Corresponding to any two points P_1 and P_2 which are in the set, every point described by $P_3 = \gamma P_1 + (1 - \gamma)P_2$ is also in the set for every $0 \leq \gamma \leq 1$. Note how this definition applies to Figure 7-9*c*; while there are some pairs of points for which this statement would be true for every γ, there are others, such as the points P_1 and P_2 which are shown, for which it is true for some γ's and not true for others. Functions of the form $\gamma P_1 + (1 - \gamma)P_2$ are known as *combinations* of P_1 and P_2, and if, in addition, $0 \leq \gamma \leq 1$, they are known as *convex combinations*. With this definition we can define a convex set more succinctly: A set is convex if and only if every convex combination of every pair of points in the set is also in the set.

We also need to define convex and concave functions (note that there is no such thing as a concave set). A convex function is one such as Figure 7-10*a*, in which a line joining any two points on the function is always on or above the function. A concave function is illustrated in Figure 7-10*b*; a line joining any two points on the function is always on or below it. Figure 7-10*c* indicates a function which is neither concave nor convex; the linear function which is illustrated in Figure 7-10*d* is both concave and convex.

There are some theorems in nonlinear programming which say that certain problems involving convex constraint sets are comparatively easy; namely, those involving the maximization of a concave objective function or the minimization of a convex objective function (which are really the same thing, since the negative of a convex function is a concave function and vice versa). Incidentally, this implies that linear programming problems are easy, which we already know. We stress the word "comparatively" above; even "easy" nonlinear programming problems are generally much harder than linear programming problems.

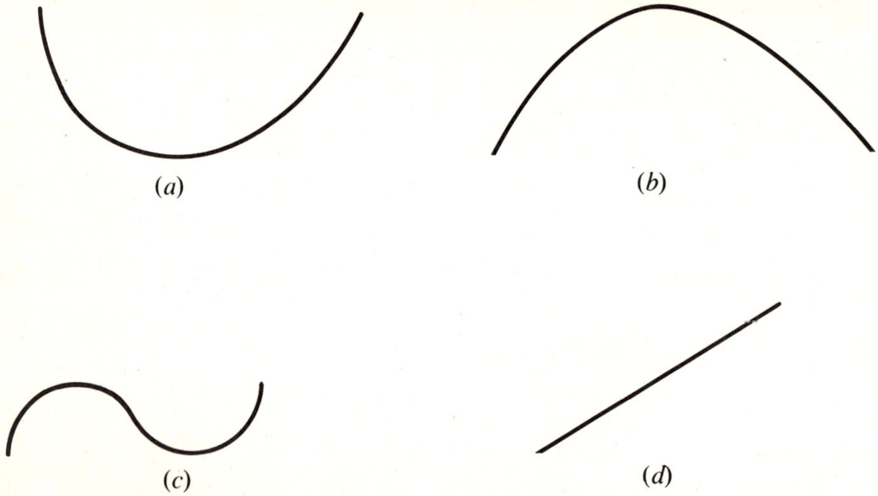

FIGURE 7-10
Concave and convex functions. (*a*) A convex function; (*b*) a concave function; (*c*) a function which is neither concave nor convex; (*d*) a function which is both concave and convex.

In Section 7.4.4 we showed why a certain problem with a concave function was difficult. Now we give another example showing how much difference convexity can make. Consider a regular linear programming problem, perhaps a diet problem, in which the cost of one of the variables has price breaks as shown in Figure 7-11. The price of this particular variable, say, x_8, is 10 dollars each for the first 100 units, and then a different amount for additional units—higher on the left (this is a scarce item; we have to pay extra if we want more than 100 units) and lower on the right (we can begin to buy it wholesale). These particular functions are called *piecewise linear functions* and constitute a special (and in some cases comparatively easy) form of nonlinear function. The one on the left is convex and the one on the right is concave. Since we are minimizing, the one on the left will turn out to be easy and the one on the right will turn out to be hard.

Let us show why the one on the left is going to be easy. Suppose the price is \$10 per unit for the first 100 units and \$15 per unit for additional units. Then we eliminate x_8 and substitute two new variables, x_9 and x_{10}, with $0 \leq x_9 \leq 100$ and $0 \leq x_{10}$. In each of the constraint functions we insert $a_{i8}(x_9 + x_{10})$ wherever $a_{i8}x_8$ appeared before; in the cost function we substitute $10x_9 + 15x_{10}$; and now we simply solve as a linear problem. Obviously the computer will not buy any of the expensive x_{10} as long as it can buy the inexpensive x_9; but when it reaches its upper limit of $x_9 = 100$ it will proceed to buy x_{10} as necessary. Now note that this same trick will not work in the case on the right of Figure 7-11, because the computer will try to buy the cheaper material

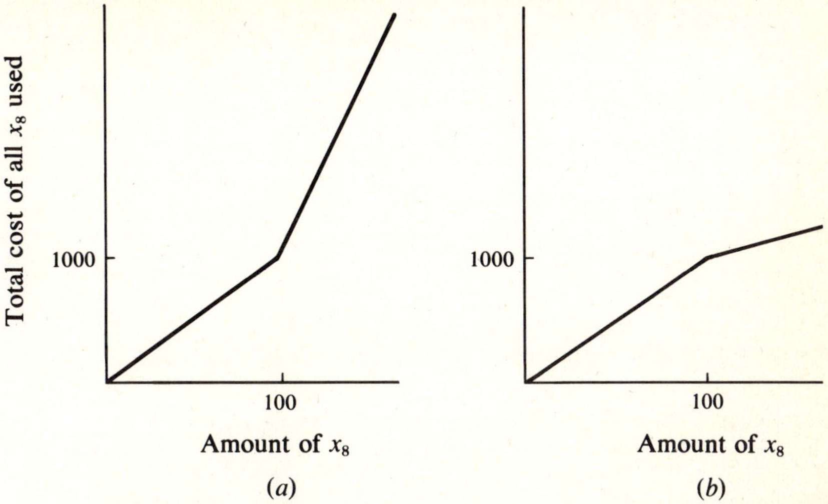

FIGURE 7-11
Piecewise linear cost functions. (*a*) Convex; (*b*) concave.

x_{10} without using any of the more expensive x_9. There is no easy way out of this, but there are ways out of it, which can be found in more advanced texts. This algorithm, and one of the algorithms for quadratic programming, are easy enough so you could understand them if you have read this far. But most nonlinear programming algorithms are more difficult, and we carry the subject no farther in this book.

Problem

7.7-1 Consider the standard linear programming problem:

Maximize: $CX = z$

Subject to: $AX \leq B$
$X \geq 0$

Call the maximum value of z achieved by solving this problem z_0.

(*a*) Let one of the b_i (an element of the vector B) vary. As it varies, z_0 may vary. Consider then the function in which b_i is the independent variable and z_0 is the dependent variable. Is this function concave or convex?

(*b*) Similarly, consider the function in which z_0 is dependent on one of the c_j's. Is this function concave or convex?

(*c*) What do these conclusions add to our knowledge of postoptimality analysis? Specifically, the "shadow prices" tell us the marginal cost of certain changes, provided those changes are small; we are now in a position to assert whether these shadow prices underestimate or overestimate if the change is large. For example, a shadow price of 7 might tell us that one additional unit of resource was worth \$7.00 but not necessarily that 10 additional units would be worth \$70.00; we are now in a position to say that it is worth \$70.00 or less than \$70.00 (not more).

8
CONCLUSION

We started this book with an examination of the assignment problem. This turns out to be a special (and simple) case of the transportation problem, which we studied next. That, in turn, is a special (and simple) case of the linear programming problem, which we then studied in some detail. Finally we indicated briefly that linear programming is a special (and simple) case of nonlinear programming, which, in turn, is a special case of mathematical programming. Mathematical programming, in turn, is a special case of the set of mathematical techniques which are useful in decision making and, in particular, in the analysis of complex systems. The name *operations research* has been given to those quantitative studies of operating systems for the purpose of aiding decision makers which often result in the use of such mathematical techniques. And the name *systems analysis* has been given to the analytic study of complex systems, which also often uses the same techniques. It is not clear that any more precise definitions of these terms would serve a useful purpose here; we mention only that you have been given here an introduction to operations research and systems analysis.

We might also indicate why we chose linear programming specifically for this introductory example rather than one of the innumerable other techniques.

It is true that linear programming is of wide applicability; but more significantly, linear programming is ideal for a book such as this because it can be mastered (at least to the extent that you have now mastered it) without the use of calculus. If you are really going to go ahead and learn more about operations research and systems analysis, you'll have to become proficient in the calculus.

On page xiii we asked you to look at the problem (2-1), presented in the six lines of equations on page 54, and added "this must appear forbidding to you and absolutely incomprehensible. By the time you get to it it will appear merely difficult. By the time you finish Chapter 2, it will appear easy. By the time you finish the book, it will appear absolutely trivial." You have now finished the book, and if we have kept faith with you, you should be able to look back at this problem and agree that it is trivial.

You should, by now, have a pretty deep understanding of the assignment problem. It turns out that you can never really understand any technical subject until you have studied the next subject beyond. In this case you have studied the next subject beyond—namely, linear programming; and through that you have gained quite a lot of understanding of the assignment problem, which is by no means as simple as you probably thought it was when you read Chapter 2. If you wanted to gain a similar depth of understanding of linear programming you would probably have to go on and study nonlinear programming, and this you probably cannot do because you do not have the necessary mathematical prerequisites.

In fact, this may be the last mathematics course that you ever study. You've spent a lot of time and effort on this one, and it is reasonable to examine at this point why you did so and what you got out of it. The first thing that comes to mind is that you learned how to solve certain problems; but let's face it: You will probably never again actually solve a linear programming problem; you will almost certainly never again solve one by hand. On the other hand, you may have to write a computer program some day, and you are certainly better qualified to do that because of your understanding of the algorithmic method and how to manipulate arrays of numbers.

Even if you never write a computer program you may some day be a decision maker faced with a computer output which may aid you in making a decision. Toward this end you should by now have gathered some sophistication as to the degree to which you should accept such computer output. Clearly there are times when the computer output is very useful; and clearly there are times when it is useless, or even counterproductive and dangerous. This is not, of course, because the computer itself makes errors—the horrible blunders that people sometimes make in believing a computer are due to inadequate models or inadequate data. You can now understand how, if a situation has been modeled linearly whereas the real situation is strongly nonlinear, the resultant computer solution may be badly in error. You also have a feeling (from your study of sensitivity analysis) to what extent errors in the data can affect the conclusions. We have seen some cases where very large errors in the data have no effect at all (for example, in the case of an inactive constraint); we have also seen cases where very minute changes in the data can have a very gross effect on the solution (see, for example, Figure 7-7). More and more in this increasingly quantitative and computer-oriented world, decision makers are going to have to make this kind of assessment. They will be presented with elaborate simulations and computations developed by smart young mathematicians; but it may well be the nonmathematical manager, whose educational level of mathematical sophistication is no greater than yours, who will have to decide just how valid these results are.

And even if you are not a decision maker, to be an alert and well-informed citizen in this modern world you must frequently make decisions similar to the above, in which you must take the proper middle course between acceptance and skepticism when faced with conclusions or recommendations which arise from the manipulation of a mathematical model. We think you are now much better able to make such decisions.

Finally, there is a still more general achievement which you should have gotten from mastering this book, and that is a better understanding of what mathematics is all about. After all, an educated person should have some feeling for what each of the major disciplines (including mathematics, the natural sciences, the social sciences, and the humanities) is all about. Mathematics has been called "the queen of the sciences"; but most people never get beyond the concept that it has something to do with numbers and formulas and is terribly forbidding. We hope that by now you have a real understanding of our assertion made on page xii that, while numbers and formulas are useful in mathematics, what it is really all about is relationships of any kind when these can be expressed quantitatively. To help convince you of this, we describe the following mathematical problem which is essentially free of any relationship to numbers.

Consider an ordinary checkerboard 8 inches by 8 inches, with sixty-four 1-inch squares. You are given a set of 31 dominoes, each 1 inch by 2 inches, so that each domino can exactly cover two adjacent squares. The problem is to place the dominoes on the checkerboard in such a way that all the squares of the checkerboard are covered, except for the ones in the northwest corner and the southeast corner. If you wish to try this, you will undoubtedly convince yourself after a few minutes that it is an extremely difficult problem and perhaps impossible. The mathematician knows that constructing a solution is not the only method of solving such a problem; another method is to prove that no solution can exist. Let us try that approach to this problem. Note that alternate squares on a checkerboard are of different colors, say, red and black. Note that each domino must cover two squares of different colors. Finally, note that the two squares which we have asked to be left uncovered are both of the same color. It follows that the problem is insoluble. This is real mathematics. Note that what numbers there were in the problem are essentially irrelevant. The proof would have applied whether the squares were 1 inch by 1 inch or some other size and whether the checkerboard were eight by eight or six by six or some other size.

We think it would be fatuous to attempt to describe here how a mathematician approaches a problem—that is what we have been doing in this entire book—but we can point out some aspects of it. In the first place, he or she attempts to extract from a real situation some precise problem which can be modeled quantitatively. In the second place the mathematician tends to imbed the given problem in a whole class of problems, since solving the whole class is often easier than solving the special problem. Thus, when faced with the 10×10 assignment problem of Figure 2-4, the mathematician goes instead to the representation of equations (2-1), which apply to any assignment problem of any size. He or she then investigates the mathematical properties of all such problems and, in particular, the mathematical properties which any solution must have. Note how the linear programming problem in general has an infinite number of feasible solutions. Dantzig's Theorem immediately reduces the number of feasible solutions which can conceivably be candidates for optimality to a finite number, and the additional theoretical knowledge which leads us to the simplex algorithm enables us to examine only a small fraction of those candidates. Thus our algorithms, which are very clever and very efficient, come about not by happenstance but by a deep sophisticated mathematical understanding of the problem and of the properties which any solution to that problem must have.

If you've gotten some feeling for this method of approaching the world and attempting to gain an understanding of it, then the effort you have expended and the pain you have undergone in mastering this book will have been well worthwhile.

INDEX